Metabolic Network Analysis

Metabolic Network Analysis

Editor

Dr. T.P. Jayadev Reddy

BIOTECH

2007

Biotech Books

Delhi - 110 035

ISBN 81-7622-169-4

Published by : **BIOTECH BOOKS**
1123/74, Tri Nagar,
DELHI – 110 035
Phone: 27382765
e-mail: biotechbooks@yahoo.co.in

Showroom : 4762-63/23, Ansari Road, Darya Ganj,
NEW DELHI - 110 002
Phone: 23245578, 23244987

Printed at : **Chawla Offset Printers**
New Delhi – 110 052

PRINTED IN INDIA

PREFACE

Metabolism is a set of chemical reactions, used by living organisms to process chemical compounds in order to take energy, extract building blocks and eliminate toxic compounds. Most of these reactions would not be executed without specialised proteins called enzymes, whose function is to catalyse these chemical reactions. Metabolism comprises the network of interactions that provide energy and building blocks for cells and organisms, a network sustaining the living and allowing it to grow and reproduce. Understanding the nature and functions of metabolic networks is imperative to biology, toxicology and medicine.

Metabolism was previously seen as a combination of distinct pathways, such as glycolysis, citrate cycle, urea cycle, amino acid biosynthesis and many others. All these pathways are connected to each other. In recent years, metabolism has started to be studied in a network approach. With the sequencing of complete genomes, it is now possible to reconstruct the network of biochemical reactions in many organisms, from bacteria to human. Information about the structure of metabolic networks can now be partially extracted and represented in graphical form using KEGG, WIT and MetaCyc. For example, the data of the Kyoto Encyclopaedia of gene and Genomes (KEGG) consists of information on interacting molecular and gene pathways.

This book examines how biological components interact and form metabolic networks, and how these networks generate cell functions corresponding to observable phenotypes. It describes how to model metabolic networks, how to determine their properties, and how to relate these to phenotypic functions. It will be a highly beneficial reference tool for researchers and scholars in all areas of life sciences.

Editor

CONTENTS

1

BIOLOGY OF CELLULAR NETWORKS

Network biology offers a quantifiable description of the networks that characterise various biological systems. Most biological characteristics arise from complex interactions between the cell's numerous constituents, such as proteins, DNA, RNA and small molecules. Therefore, a key challenge for biology in the twenty-first century is to understand the structure and the dynamics of the complex intercellular web of interactions that contribute to the structure and function of a living cell. The development of high-throughput data-collection techniques, as epitomised by the widespread use of microarrays, allows for the simultaneous interrogation of the status of a cell's components at any given time.

In turn, new technology platforms, such as PROTEIN CHIPS or semi-automated YEAST TWO-HYBRID SCREENS, help to determine how and when these molecules interact with each other. Various types of interaction webs, or networks, (including protein-protein interaction,metabolic, signalling and transcription-regulatory networks) emerge from the sum of these interactions. None of these networks are independent, instead they form a 'network of networks' that is responsible for the behaviour of the cell.

A major challenge of contemporary biology is to embark on an integrated theoretical and experimental programme to map out, understand and model in quantifiable terms the topological and dynamic properties of the various networks that control the behaviour of the cell. Help along the way is provided by the rapidly developing theory of complex networks that, in the past few years, has made advances towards uncovering the organising principles that govern the formation and evolution of various complex technological and social networks.

The architectural features of molecular interaction networks within a cell are shared to a large degree by other complex systems, such as the Internet, computer chips and society. This unexpected universality indicates that similar laws may govern most complex networks in nature, which allows the expertise from large and well-mapped non-biological systems to be used to characterise the intricate interwoven relationships that govern cellular functions.

The quantifiable tools of network theory offer unforeseen possibilities to understand the cell's internal organisation and evolution. The emerging results are forcing the realisation that, notwithstanding the importance of individual molecules, cellular function is a contextual attribute of strict and quantifiable patterns of interactions between the myriad of cellular constituents. Although uncovering the generic organising principles of cellular networks is fundamental to our understanding of the cell as a system, it also needs to develop relevance for the experimental biologist, helping to elucidate the role of individual molecules in various cellular processes.

Therefore, explore the specific biological details and the evolutionary origins that contribute to the formation of cellular networks, and the impact of the network structure on experimentally observable function and behavioural features.

Network Measures

Here define the most basic network measures that allow us to compare and characterise different complex networks. The most elementary characteristic of a node is its degree (or connectivity), k which tells us how many links the node has to other nodes.

For example, in the undirected network shown in part a of the figure 1, node A has degree $k = 5$. In networks in which each link has a selected direction (figure 1, part b) there is an incoming degree, k_{in}, which denotes the number of links that point to a node, and an outgoing degree, k_{out}, which denotes the number of links that start from it.

For example, node A in part b of the figure 1 has $k_{in} = 4$ and $k_{out} = 1$. An undirected network with N nodes and L links is characterised by an average degree $\langle k \rangle = 2L/N$ (where $\langle\rangle$ denotes the average).

Degree distribution: The degree distribution, $P(k)$, gives the probability that a selected node has exactly k links. $P(k)$ is obtained by counting the number o f nodes N(k) with $k = 1$, 2... links and dividing by the total number of nodes N. The degree distribution allows us to distinguish between different classes of networks. For example, a peaked degree distribution, as seen in a random network, indicates that the system has a characteristic degree and that there are no highly connected nodes (which are also known as hubs). By contrast, a power-law degree distribution indicates that a few hubs hold together numerous small nodes.

Scale-free networks and the degree exponent: Most biological networks are scale-free,which means that their degree distribution approximates a power law, $P(k) \sim k^{-\gamma}$, where γ is the degree exponent and ~ indicates 'proportional to'. The value of γ determines many properties of the system. The smaller the value of γ, the more important the role of the hubs is in the network. Whereas for $\gamma>3$ the hubs are not relevant, for $2> \gamma >3$ there is a hierarchy of hubs,with the most connected hub being in contact with a small fraction of all nodes, and for $\gamma = 2$ a hub-and-spoke network emerges,with the largest hub being in contact with a large fraction of all nodes.

In general, the unusual properties of scale-free networks are valid only for $\gamma<3$, when the dispersion of the $P(k)$ distribution,which is defined as $\sigma^2 = \langle k^2 \rangle - \langle k \rangle^2$, increases with the number of nodes (that is, σ diverges), resulting in a series of unexpected features, such as a high degree of robustness against accidental node failures. For $\gamma > 3$, however, most unusual features are absent, and in many respects the scale-free network behaves like a random one.

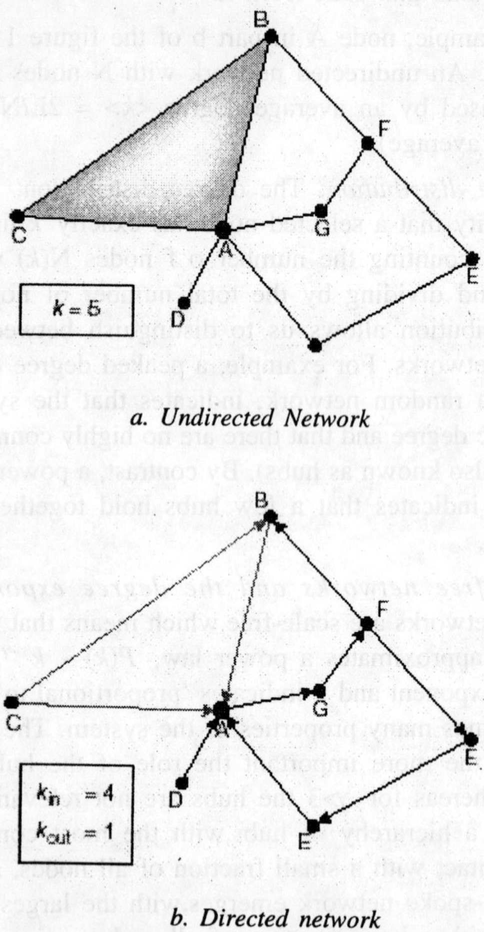

Figure 1. Undirected network and Directed network

Shortest path and mean path length: Distance in networks is measured with the path length, which tells us how many links we need to pass through to travel between two nodes. As there are many alternative paths between two nodes, the shortest path—the path with the smallest number of links between the selected nodes—has a special role. In directed networks, the distance ℓ_{AB} from node A to node B is often different from the distance ℓ_{BA} from B to A. For example, in part b of the figure 1, $\ell_{BA} = 1$, whereas $\ell_{AB} = 3$. Often there is no direct path between two nodes. As shown in part b of the figure 1, although there is a path from C to A, there is no path from A to C. The mean path length, $<\ell>$, represents the average over the shortest paths between all pairs of nodes and offers a measure of a network's overall navigability.

Clustering coefficient: In many networks, if node A is connected to B, and B is connected to C, then it is highly probable that A also has a direct link to C. This phenomenon can be quantified using the clustering coefficient $C_I = 2n_I/k(k-1)$, where n_I is the number of links connecting the k_I neighbours of node I to each other. In other words, C_I gives the number of 'triangles' that go through node I, whereas $k_I(k_I -1)/2$ is the total number of triangles that could pass through node I, should all of node I's neighbours be connected to each other. For example, only one pair of node A's five neighbours in part a of the figure 1 are linked together (B and C), which gives nA = 1 and C_A = 2/20.

By contrast, none of node F's neighbours link to each other, giving $C_F = 0$. The average clustering coefficient, $<C>$, characterises the overall tendency of nodes to form clusters or groups. An important measure of the network's structure is the function C(k), which is defined as the average clustering coefficient of all nodes with k links. For many real networks $C(k) \sim k^{-1}$, which is an indication of a network's hierarchical character. The average degree $<k>$, average path length $<\ell>$ and average clustering coefficient $<C>$ depend on the number of nodes and links (N and L) in the network. By contrast, the P(k) and C(k) functions are independent of the network's size and they therefore capture a network's generic features, which allows them to be used to classify various networks.

Network Models

Network models are crucial for shaping our understanding of complex networks and help to explain the origin of observed network characteristics. There are three models that had a direct impact on our understanding of biological networks.

Random Networks

The Erdös–Rényi (ER) model of a random network (figure 2, part A) starts with N nodes and connects each pair of nodes with probability *p*, which creates a graph with approximately $pN(N-1)/2$ randomly placed links (figure 2, part Aa). The node degrees follow a Poisson distribution (figure 2, part Ab), which indicates that most nodes have approximately the same number of links (close to the average degree $<k>$). The tail (high *k* region) of the degree distribution $P(k)$ decreases exponentially, which indicates that nodes that significantly deviate from the average are extremely rare.

The clustering coefficient is independent of a node's degree, so $C(k)$ appears as a horizontal line if plotted as a function of *k* (figure 2, part Ac). The mean path length is proportional to the logarithm of the network size, $l \sim \log N$, which indicates that it is characterised by the small-world property.

Scale-free Networks

Scale-free networks (figure 2, part B) are characterised by a power-law degree distribution; the probability that a node has k links follows $P(k) \sim k^{-\gamma}$, where γ is the degree exponent. The probability that a node is highly connected is statistically more significant than in a random graph, the network's properties often being determined by a relatively small number of highly connected nodes that are known as hubs (figure 2, part Ba). In the Barabási–Albert model of a scale-free network, at each time point a node with M links is added to the network, which connects to an already existing node I with probability $\Pi_I = k_I/$

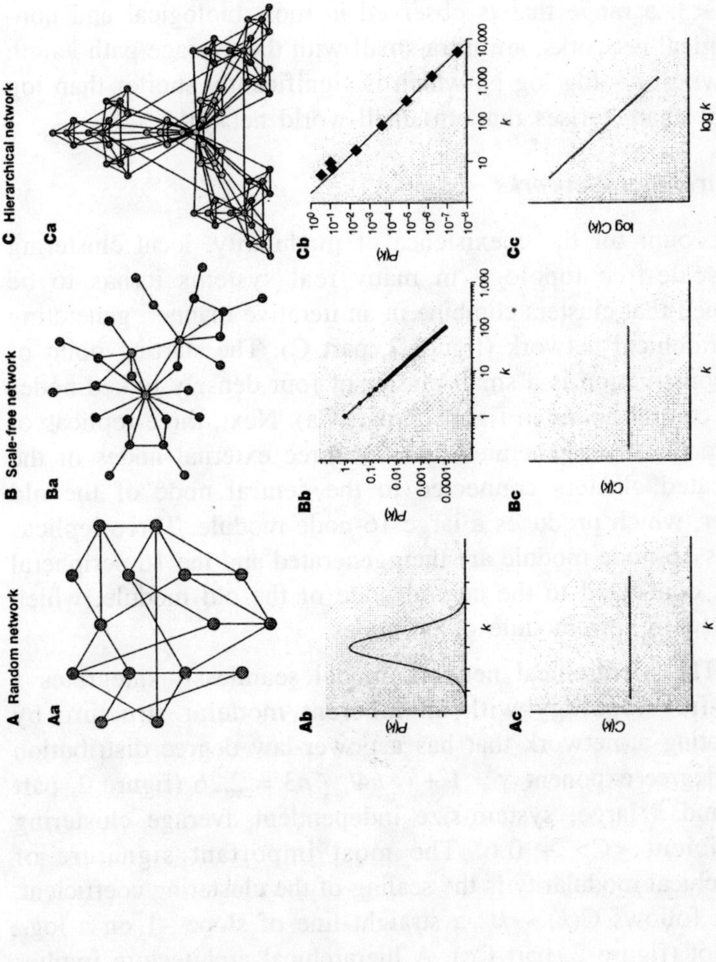

Figure 2. Network models

$\Sigma_J k_J$, where k_I is the degree of node I (Fig. 4) and J is the index denoting the sum over network nodes. The network that is generated by this growth process has a power-law degree distribution that is characterised by the degree exponent $\gamma = 3$. Such distributions are seen as a straight line on a log–log plot (figure 2, part Bb).

The network that is created by the Barabási–Albert model does not have an inherent modularity, so $C(k)$ is independent of k (figure 2, part Bc). Scale-free networks with degree exponents $2< \gamma <3$, a range that is observed in most biological and non-biological networks, are ultra-small with the average path length following $\ell \sim \log \log N$, which is significantly shorter than log N that characterises random small-world networks.

Hierarchical Networks

To account for the coexistence of modularity, local clustering and scale-free topology in many real systems it has to be assumed that clusters combine in an iterative manner, generating a hierarchical network (figure 2, part C). The starting point of this construction is a small cluster of four densely linked nodes (four central nodes in figure 2, part Ca). Next, three replicas of this module are generated and the three external nodes of the replicated clusters connected to the central node of the old cluster, which produces a large 16-node module. Three replicas of this 16-node module are then generated and the 16 peripheral nodes connected to the central node of the old module, which produces a new module of 64 nodes.

The hierarchical network model seamlessly integrates a scale-free topology with an inherent modular structure by generating a network that has a power-law degree distribution with degree exponent $\gamma = 1 + \ell n4/\ \ell n3 = 2.26$ (figure 2, part Cb) and a large, system-size independent average clustering coefficient $\langle C \rangle \sim 0.6$. The most important signature of hierarchical modularity is the scaling of the clustering coefficient, which follows $C(k) \sim k^{-1}$ a straight line of slope –1 on a log–log plot (figure 2, part Cc). A hierarchical architecture implies that sparsely connected nodes are part of highly clustered areas, with communication between the different highly clustered neighbourhoods being maintained by a few hubs (figure 2, part Ca).

NETWORK NOMENCLATURE

The behaviour of most complex systems, from the cell to the Internet, emerges from the orchestrated activity of many components that interact with each other through pairwise interactions. At a highly abstract level, the components can be reduced to a series of nodes that are connected to each other by links,with each link representing the interactions between two components. The nodes and links together form a network, or, in more formal mathematical language, a graph.

Establishing the identity of various cellular networks is not trivial. Physical interactions between molecules, such as protein-protein, protein-nucleic-acid and protein-metabolite interactions, can easily be conceptualised using the node-link nomenclature. Nevertheless, more complex functional interactions can also be considered within this representation. For example, mallmolecule substrates can be envisioned as the nodes of a metabolic network and the links as the enzyme-catalysed reactions that transform one metabolite into another (Fig 3a-c).

To study the network characteristics of the metabolism a graph theoretic description needs to be established. Figure 3 shows the theoretic description for a simple pathway (catalysed by Mg^{2+}-dependant enzymes). In this figure, (a) In the most abstract approach (b) all interacting metabolites are considered equally. The links between nodes represent reactions that interconvert one substrate into another. For many biological applications it is useful to ignore co-factors, such as the highenergy- phosphate donor ATP, which results in a second type of mapping (c) that connects only the main source metabolites to the main products. (d) The degree distribution, $P(k)$ of the metabolic network illustrates its scale-free topology.

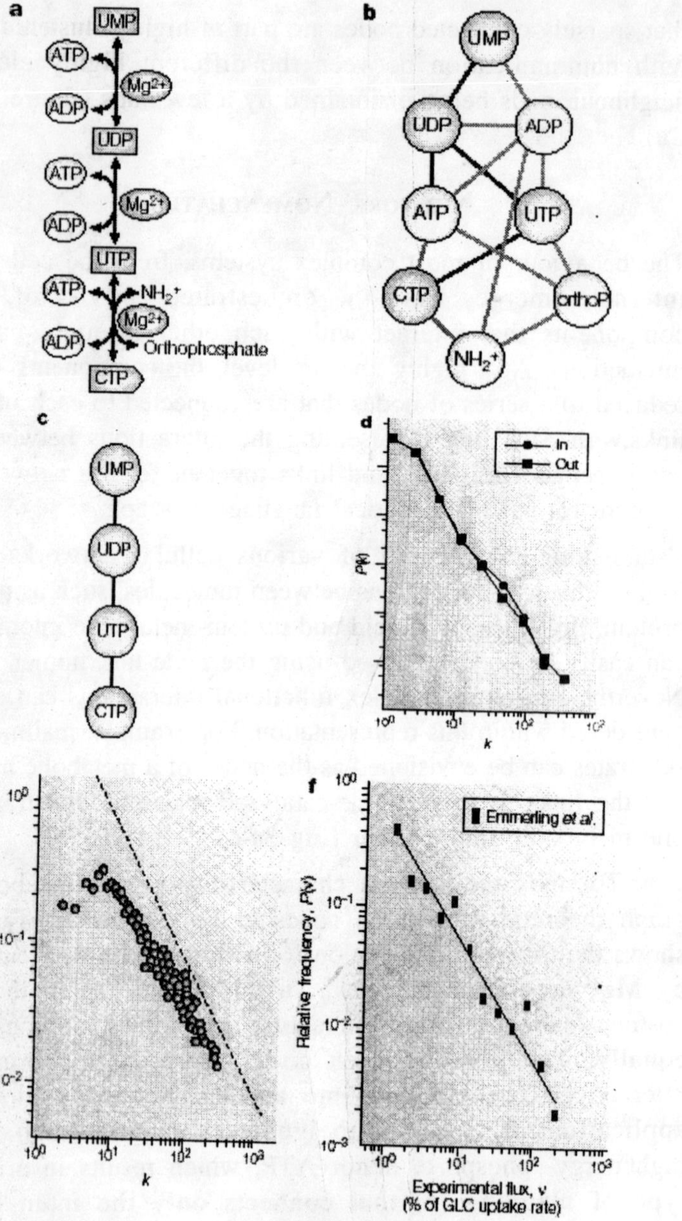

Figure 3. Characterizing metabolic networks

(e) The scaling of the clustering coefficient C(*k*) with the degree *k* illustrates the hierarchical architecture of metabolism (The data shown in d and e represent an average over 43 organisms,). (f) The flux distribution in the central metabolism of Escherichia coli follows a power law, which indicates that most reactions have small metabolic flux, whereas a few reactions, with high fluxes, carry most of the metabolic activity. On all three plots the axis is logarithmic and a straight line on such log-log plots indicates a power-law scaling. CTP, cytidine triphosphate; GLC, aldo-hexose glucose; UDP, uridine diphosphate; UMP, uridine monophosphate; UTP, uridine triphosphate.

Depending on the nature of the interactions, networks can be directed or undirected. In directed networks, the interaction between any two nodes has a well-defined direction,which represents, for example, the direction of material flow from a substrate to a product in a metabolic reaction, or the direction of information flow from a transcription factor to the gene that it regulates. In undirected networks, the links do not have an assigned direction. For example, in protein interaction networks a link represents a mutual binding relationship: if protein A binds to protein B, then protein B also binds to protein A.

Features of Cellular Networks

Probably the most important discovery of network theory was the realisation that despite the remarkable diversity of networks in nature, their architecture is governed by a few simple principles that are common to most networks of major scientific and technological interest. For decades graph theory—the field of mathematics that deals with the mathematical foundations of networks—modelled complex networks either as regular objects, such as a square or a diamond lattice, or as completely random network.

This approach was rooted in the influential work of two mathematicians, Paul Erdös, and Alfréd Rényi,who initiated the study of the mathematical properties of random networks. Their much-investigated random network model assumes that a fixed number of nodes are connected randomly to each other. The most remarkable property of the model is its 'democratic' or uniform character, characterising the degree,or connectivity (k), of the individual nodes. Because, in the model, the links are placed randomly among the nodes, it is expected that some nodes collect only a few links whereas others collect many more.

In a random network, the nodes degrees follow a Poisson distribution, which indicates that most nodes have roughly the same number of links, approximately equal to the network's average degree,$<k>$ (where $<>$ denotes the average); nodes that have significantly more or less links than $<k>$ are absent or very rare. Despite its elegance, a series of recent findings indicate that the random network model cannot explain the topological properties of real networks. The deviations from the random model have several key signatures, the most striking being the finding that, in contrast to the Poisson degree distribution, for many social and technological networks the number of nodes with a given degree follows a power law. That is, the probability that a chosen node has exactly k links follows $P(k) \sim k^{-\gamma}$, where γ is the degree exponent, with its value for most networks being between 2 and 3.

Networks that are characterised by a power-law degree distribution are highly non-uniform, most of the nodes have only a few links. A few nodes with a very large number of links, which are often called hubs, hold these nodes together. Networks with a power degree distribution are called scale-free, a name that is rooted in statistical physics literature. It indicates the absence of a typical node in the network (one that could be used to characterise the rest of the nodes). This is in strong contrast to random networks, for which the degree of all nodes is in the vicinity of the average degree, which could be considered typical. However, scale-free networks could easily be called scale-rich as well, as their main feature is the coexistence of nodes of widely different degrees (scales), from nodes with one or two links to major hubs.

Scale-free Cellular Networks

Most networks within the cell approximate a scale-free topology. The first evidence came from the analysis of metabolism, in which the nodes are metabolites and the links represent enzyme-catalysed biochemical reactions (Fig 3). As many of the reactions are irreversible, metabolic networks are directed. So, for each metabolite an 'in' and an 'out' degree can be assigned that denotes the number of reactions that produce or consume it, respectively.

The analysis of the metabolic networks of 43 different organisms from all three domains of life (eukaryotes, bacteria, and archaea) indicates that the cellular metabolism has a scale-free topology, in which most metabolic substrates participate in only one or two reactions, but a few, such as pyruvate or coenzyme A, participate in dozens and function as metabolic hubs. As for direct physical interactions, several recent publications indicate that protein-protein interactions in diverse eukaryotic species also have the features of a scale-free network.

Examples of scale-free organisation include genetic regulatory networks, in which the nodes are individual genes and the links are derived from the expression correlations that are based on microarray data or protein domain networks that are constructed on the basis of protein domain interactions. However, not all networks within the cell are scale-free. For example, the transcription regulatory networks of S. cerevisiae and Escherichia coli offer an interesting example of mixed scale-free and exponential characteristics. Indeed, the distribution that captures how many different genes a transcription factor interacts with follows a power law, which is a signature of a scale-free network.

This indicates that most transcription factors regulate only - few genes, but a few general transcription factors interact with many genes. However, the incoming degree distribution,which tells us how many different transcription factors interact with a given gene, is best approximated by an exponential, which indicates that most genes are regulated by one to three transcription factors. So, the key message is the recognition that cellular networks have a disproportionate number of highly connected nodes. Although the mathematical definition of a scale-free network requires us to establish that the degree distribution follows a power law, which is difficult in networks with too few nodes, the presence of hubs seems to be a general feature of all cellular networks, from regulatory webs to the p53 module. These hubs fundamentally determine the network's behaviour.

Small-World Effect and Assortativity

A common feature of all complex networks is that any two nodes can be connected with a path of a few links only. This 'small-world effect', which was originally observed in a social study, has been subsequently shown in several systems, from neural networks to the World Wide Web. Although the small-world effect is a property of random networks, scale-free networks are ultra small,—their path length is much shorter than predicted by the small-world effect.

Within the cell, this ultrasmall—world effect was first documented for metabolism, where paths of only three to four reactions can link most pairs of metabolites. This short path length indicates that local perturbations in metabolite concentrations could reach the whole network very quickly. Interestingly, the evolutionarily reduced metabolic network of a parasitic bacterium has the same mean path length as the highly developed network of a large multicellular organism,which indicates that there are evolutionary mechanisms that have maintained the average path length during evolution.

Disassortative nature of cellular networks indicates, for example, that, in protein interaction networks, highly connected nodes (hubs) avoid linking directly to each other and instead connect to proteins with only a few interactions. In contrast to the assortative nature of social networks, in which wellconnected people tend to know each other, disassortativity seems to be a property of all biological (metabolic, protein interaction) and technological (World Wide Web, Internet) networks. Although the small—and ultra-small-world property of complex networks is mathematically well understood the origin of disassortativity in cellular networks remains unexplained.

Evolutionary origin of scale-free networks. The ubiquity of scale-free networks and hubs in technological, biological and social systems requires an explanation. It has emerged that two fundamental processes have a key role in the development of real networks. First, most networks are the result of a growth process, during which new nodes join the system over an extended time period. This is the case for the World Wide Web, which has grown from 1 to more than 3-billion web pages over a 10-year period. Second, nodes prefer to connect to nodes that already have many links, a process that is known as preferential attachment. For example, on the World Wide Web we are more familiar with the highly connected web pages, and therefore are more likely to link to them. Growth and preferential attachment are jointly responsible for the emergence of the scale-free property in complex networks (Fig. 4a).

Indeed, if a node has many links, new nodes will tend to connect to it with a higher probability. This node will therefore gain new links at a higher rate than its less connected peers and will turn into a hub. Growth and preferential attachment have a common origin in protein networks that is probably rooted in gene duplication. Duplicated genes produce identical proteins that interact with the same protein partners (Fig. 4).

Therefore, each protein that is in contact with a duplicated protein gains an extra link. Highly connected proteins have a natural advantage: it is not that they are more (or less) likely to be duplicated, but they are more likely to have a link to a duplicated protein than their weakly connected cousins, and therefore they are more likely to gain new links if a randomly selected protein is duplicated. This bias represents a subtle version of preferential attachment.

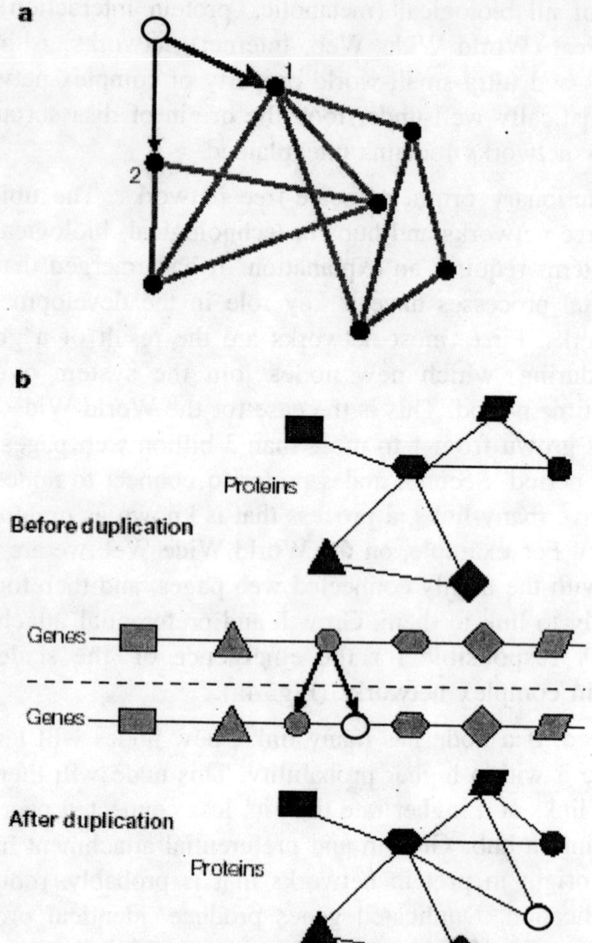

Figure 4. The origin of the scale-free topology and hubs in biological networks. (a)Growth and preferential attachment are jointly responsible for the emergence of the scale-free property in complex networks. (b) a small protein interaction network (Black) and the genes that encode the proteins (Grey).

The most important feature of this explanation is that it traces the origin of the scale-free topology back to a well-known biological mechanism—gene duplication. Although the role of gene duplication has been shown only for protein interaction networks, it probably explains, with appropriate adjustments, the emergence of the scale-free features in the regulatory and metabolic networks as well. It should be noted that, although the models show beyond doubt that gene duplication can lead to a scale-free topology, there is no direct proof that this mechanism is the only one, or the one that generates the observed power laws in cellular networks.

However, as gene duplication is a major engineer of the genomic landscape, it is likely to be a key mechanism for generating the scale-free topology. Two further results offer direct evidence that network growth is responsible for the observed topological features. The scale-free model predicts that the nodes that appeared early in the history of the network are the most connected ones. Indeed, an inspection of the metabolic hubs indicates that the remnants of the RNA world, such as coenzyme A, NAD and GTP, are among the most connected substrates of the metabolic network, as are elements of some of the most ancient metabolic pathways, such as glycolysis and the tricarboxylic acid cycle.

The origin of the scale-free topology in complex networks can be reduced to two basic mechanisms: growth and preferential attachment. Growth means that the network emerges through the subsequent addition of new nodes, such as the new white node that is added to the network that is shown in part a. Preferential attachment means that new nodes prefer to link to more connected nodes.

For example, the probability that the white node will connect to node 1 is twice as large as connecting to node 2, as the degree of node 1 (k_1=4) is twice the degree of node 2 (k_2=2). Growth and preferential attachment generate hubs through a 'rich-gets-richer' mechanism: the more connected a node is, the more likely it is that new nodes will link to it, which allows the highly connected nodes to acquire new links faster than their less connected peers. In protein interaction networks, scale-free topology seems to have its origin in gene duplication. Part b shows a small protein interaction network (black) and the genes that encode the proteins (grey). When cells divide, occasionally one or several genes are copied twice into the offspring's genome (illustrated by the ash and white circles). This induces growth in the protein interaction network because now we have an extra gene that encodes a new protein (white circle). The new protein has the same structure as the old one, so they both interact with the same proteins.

Ultimately, the proteins that interacted with the original duplicated protein will each gain a new interaction to the new protein. Therefore proteins with a large number of interactions tend to gain links more often, as it is more likely that they interact with the protein that has been duplicated. This is a mechanism that generates preferential attachment in cellular networks. Indeed, in the example that is shown in part b it does not matter which gene is duplicated, the most connected central protein (hub) gains one interaction. In contrast, the square, which has only one link, gains a new link only if the hub is duplicated.

In the context of the protein interaction networks, cross-genome comparisons have found that, on average, the evolutionarily older proteins have more links to other proteins than their younger counterparts. This offers direct empirical evidence for preferential attachment.

Subgraphs, Motifs and Motif Clusters

Cellular functions are likely to be carried out in a highly modular manner. In general, modularity refers to a group of physically or functionally linked molecules (nodes) that work together to achieve a (relatively) distinct function. Modules are seen in many systems, for example, circles of friends in social networks or websites that are devoted to similar topics on the World Wide Web.

Whereas the scale-free and hierarchical features of complex networks emphasise the organising principles that determine the network's large-scale structure, an alternative approach starts from the bottom and looks for highly representative patterns of interactions that characterise specific networks.

A connected subgraph represents a subset of nodes that are connected to each other in a specific wiring diagramme. For example, in part a of the figure 5, four nodes that form a little square represent a subgraph of a square lattice. Networks with a more intricate wiring diagramme can have various different subgraphs. For example, in part Aof the figure 1, nodes A,B and C form a triangle subgraph, whereas A,B, F and G form a square subgraph.Examples of different potential subgraphs that are present in undirected networks are shown in part b of the figure (a directed network is shown in part c). It should be noted that the number of distinct subgraphs grows exponentially with an increasing number of nodes.

Not all subgraphs occur with equal frequency. Indeed, the square lattice (figure 5, part a) contains many squares,but no triangles. In a complex network with an apparently random wiring diagramme all subgraphs—from triangles to squares and pentagons and so on—are present. However, some subgraphs,which are known as motifs, are over represented as compared to a randomised version of the same network. For example, the directed triangle motif that is known as the feed-forward loop (figure 5, top of part c) emerges in both transcription-regulatory and neural networks,whereas four-node feedback loops (figure 5, middle of part c) represent characteristic motifs in electric circuits but not in biological systems.

To identify the motifs that characterise a given network, all subgraphs of *n* nodes in the network are determined.Next, the network is randomised while keeping the number of nodes, links and the degree distribution unchanged. Subgraphs that occur significantly more frequently in the real network, as compared to randomised one, are designated to be the motifs.

Motifs are elementary units of cellular networks. Motifs and subgraphs that occur in a given network are not independent of each other. In part d of the figure 5, all of the 209 bi-fan motifs (a motif with 4 nodes) that are found in the Escherichia coli transcription-regulatory network are shown simultaneously. As the figure shows, 208 of the 209 bi-fan motifs form two extended motif clusters and only one motif remains isolated (bottom left corner). Such clustering of motifs into motif clusters seems to be a general property of all real networks. The different shapes of the nodes illustrate their functional classification.

Similarly, in many complex engineered systems, from a modern aircraft to a computer chip, a highly modular structure is a fundamental design attribute. Biology is full of examples of modularity. Relatively invariant protein-protein and protein-RNA complexes (physical modules) are at the core of many basic biological functions, from nucleic-acid synthesis to protein degradation.

Similarly, temporally coregulated groups of molecules are known to govern various stages of the cell cycle, or to convey extracellular signals in bacterial chemotaxis or the yeast pheromone response pathway. In fact, most molecules in a cell are either part of an intracellular complex with modular activity, such as the ribosome, or they participate in an extended (functional) module as a temporally regulated element of a relatively distinct process (for example, signal amplification in a signalling pathway). To address the modularity of networks, tools and measures need to be developed that will allow us not only to establish if a network is modular, but also to explicitly identify the modules and their relationships in a given network.

High Clustering

In a network representation, a module (or cluster) appears as a highly interconnected group of nodes. Each module can be reduced to a set of triangles; a high density of trianglesis reflected by the clustering coefficient, C, the signature of a network's potential modularity. In the absence of modularity,

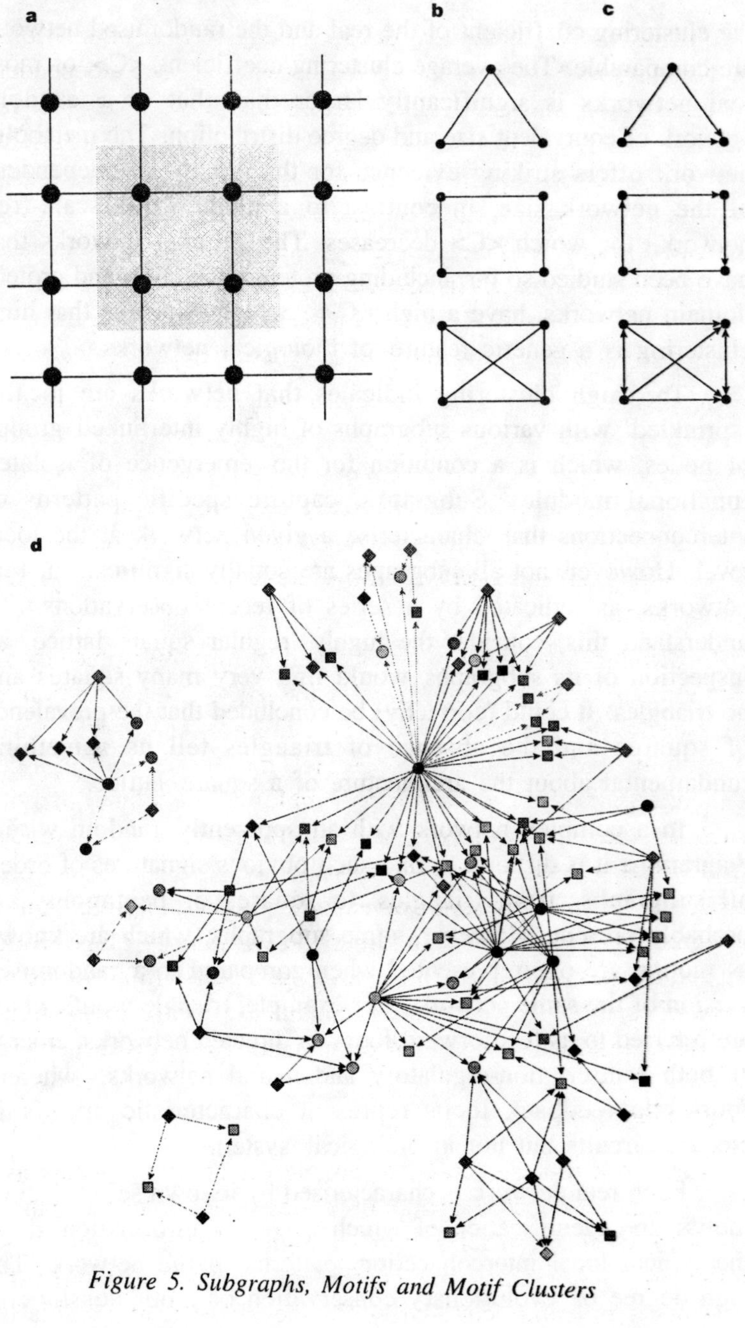

Figure 5. Subgraphs, Motifs and Motif Clusters

the clustering coefficient of the real and the randomised network are comparable. The average clustering coefficient, <C>, of most real networks is significantly larger than that of a random network of equivalent size and degree distribution. The metabolic network offers striking evidence for this: <C > is independent of the network size, in contrast to a module-free scale-free network, for which <C> decreases. The cellular networks that have been studied so far, including protein interaction and protein domain networks, have a high <C >, which indicates that high clustering is a generic feature of biological networks.

The high clustering indicates that networks are locally 'sprinkled' with various subgraphs of highly interlinked groups of nodes, which is a condition for the emergence of isolated functional modules. Subgraphs capture specific patterns of interconnections that characterise a given network at the local level. However, not all subgraphs are equally significant in real networks, as indicated by a series of recent observations. To understand this, consider the highly regular square lattice: an inspection of its subgraphs would find very many squares and no triangles. It could (correctly) be concluded that the prevalence of squares and the absence of triangles tell us something fundamental about the architecture of a square lattice.

In a complex network with an apparently random wiring diagramme it is difficult to find such obvious signatures of order: all subgraphs, from triangles to squares or pentagons, are probably present. However, some subgraphs, which are known as motifs, are overrepresented when compared to a randomised version of the same network,. For example, triangle motifs,which are referred to as feedforward loops in directed networks, emerge in both transcription-regulatory and neural networks, whereas four-node feedback loops represent characteristic motifs in electric circuits but not in biological systems.

Each real network is characterised by its own set of distinct motifs, the identification of which provides information about the typical local interconnection patterns in the network. The high degree of evolutionary conservation of motif constituents within the yeast protein interaction network and the convergent evolution that is seen in the transcriptionregulatory network of diverse species towards the same motif types, further indicate that motifs are indeed of direct biological relevance.

As the molecular components of a specific motif often interact with nodes that are outside the motif, how the different motifs interact with each other needs to be addressed. Empirical observations indicate that specific motif types aggregate to form large motif clusters. For example, in the E. coli transcriptionregulatory network, most motifs overlap, generating distinct homologous motif clusters, in which the specific motifs are no longer clearly separable. As motifs are present in all of the real networks that have been examined so far, it is likely that the aggregation of motifs into motif clusters is a general property of most real networks.

As the number of distinct subgraphs grows exponentially with the number of nodes that are in the subgraph, the study of larger motifs is combinatorially unfeasible.An alternative approach involves identifying groups of highly interconnected nodes, or modules, directly from the graph's topology and correlating these topological entities with their potential functional role. Module identification is complicated by the fact that at face value the scale-free property and modularity seem to be contradictory.

Modules by definition imply that there are groups of nodes that are relatively isolated from the rest of the system. However, in a scale-free network hubs are in contact with a high fraction of nodes, which makes the existence of relatively isolated modules unlikely. Clustering and hubs naturally coexist, however, which indicates that topological modules are not independent, but combine to form a hierarchical network.

An example of such a hierarchical network is shown in figure 2; this network is simultaneously scale-free and has a high clustering coefficient that is independent of system size. The network is made of many small, highly integrated 4-node modules that are assembled into larger 16-node modules, each of which combines in a hierarchical fashion into even larger 64-node modules. The quantifiable signature of hierarchical modularity is the dependence of the clustering coefficient on the degree of a node,which follows $C(k) \sim k^{-1}$. This indicates that nodes with only a few links have a high C and belong to highly interconnected small modules.

By contrast, the highly connected hubs have a low C, with their role being to link different, and otherwise not communicating,modules. It should be noted that the random and scale-free models that are shown in figure 2 do not have a hierarchical topology, because $C(k)$ is independent of k in their case. This is not surprising, as their construction does not contain elements that would favour the emergence of modules.

Topological and Functional Modules

Signatures of hierarchical modularity are present in all cellular networks that have been investigated so far, ranging from metabolic to protein-protein interaction, and regulatory networks. If there are clearly separated modules in the system, most clustering methods can identify them. Indeed, several methods have recently been introduced to identify modules in various networks, using either the network's topological description, or combining the topology with integrated functional genomics data.

Different methods predict different boundaries between modules that are not sharply separated. This ambiguity is not only a limitation of the present clustering methods, but it is a consequence of the network's hierarchical modularity. The hierarchical modularity indicates that modules do not have a characteristic size: the network is as likely to be partitioned into a set of clusters of 10-20 components (metabolites, genes) as into fewer, but larger modules. At present there are no objective mathematical criteria for deciding that one partition is better than another.

Indeed, in most of the present clustering algorithms some internal parameter controls the typical size of the uncovered modules, and changing the parameter results in a different set of larger or smaller modules. From a mathematical perspective it does indeed indicate that looking for a set of unique modules is an ill-defined problem. An easy solution, however, is to avoid seeking a breakdown into an absolute set of modules, but rather to visualise the hierarchical relationship between modules of different sizes. The identification of the groups of molecules of various sizes that together carry out a specific cellular function is a key issue in network biology, and one that is likely to witness much progress in the near future.

Robustness of Cellular Network

A key feature of many complex systems is their robustness, which refers to the system's ability to respond to changes in the external conditions or internal organisation while maintaining relatively normal behaviour. To understand the cell's functional organisation, insights into the interplay between the network structure and robustness, as well as their joint evolutionary origins, are needed. Topological robustness. Intuition tells us that disabling a substantial number of nodes will result in an inevitable functional disintegration of a network. This is certainly true for a random network: if a critical fraction of nodes is removed, a phase transition is observed, breaking the network into tiny, non-communicating islands of nodes.

Complex systems, from the cell to the Internet, can be amazingly resilient against component failure, withstanding even the incapacitation of many of their individual components and many changes in external conditions. Topology has an important role in generating this topological robustness. Scale-free networks do not have a critical threshold for disintegration—they are amazingly robust against accidental failures: even if 80% of randomly selected nodes fail, the remaining 20% still form a compact cluster with a path connecting any two nodes.

This is because random failure affects mainly the numerous small degree nodes, the absence of which doesn't disrupt the network's integrity. This reliance on hubs, on the other hand, induces a so-called attack vulnerability the removal of a few key hubs splinters the system into small isolated node clusters. This double-edged feature of scale-free networks indicates that there is a strong relationship between the hub status of a molecule (for example, its number of links) and its role in maintaining the viability and/or growth of a cell.

Deletion analyses indicate that in S. cerevisiae only ~10% of the proteins with less than 5 links are essential, but this fraction increases to over 60% for proteins with more than 15 interactions, which indicates that the protein's degree of connectedness has an important role in determining its deletion phenotype. Furthermore, only ~ 18.7% of S. cerevisiae genes (~14.4% in E. coli) are lethal when deleted individually-, and the simultaneous deletion of many E. coli genes is without substantial phenotypic effect,. These results are in line with the expectation that many lightly connected nodes in a scale-free network do not have a major effect on the network's integrity.

The importance of hubs is further corroborated by their evolutionary conservation: highly interacting S. cerevisiae proteins have a smaller evolutionary distance to their orthologues in Caenorhabditis elegans and are more likely to have orthologues in higher organisms.

A complete understanding of network robustness requires that the functional and dynamic changes that are caused by perturbations are explored. In a cellular network, each node has a slightly different biological function and therefore the effect of a perturbation cannot depend on the node's degree only. This is well illustrated by the finding that experimentally identified protein complexes tend to be composed of uniformly essential or non-essential molecules. This indicates that the functional role (dispensability) of the whole complex determines the deletion phenotype of the individual proteins.

The functional and dynamical robustness of cellular networks is supported by recent results that indicate that several relatively well-delineated extended modules are robust to many varied perturbations. Dynamical and functional robustness are also selective: whereas some important parameters remain unchanged under perturbations, others vary widely.

Few important themes have emerged. First, it is increasingly accepted that adaptation and robustness are inherent network properties, and not a result of the fine-tuning of a component's characteristics. Second, robustness is inevitably accompanied by vulnerabilities: although many cellular networks are well adapted to compensate for the most common perturbations, they collapse when wellselected network components are disrupted. Third, the ability of a module to evolve also has a key role in developing or limiting robustness. Indeed, evolutionarily 'frozen' modules that are responsible for key cellular functions, such as nucleic-acid synthesis, might be less able to withstand uncommon errors, such as the inactivation of two molecules within the same functional module. Finally, modularity and robustness are presumably considerably quite intertwined,with the weak communication between modules probably limiting the effects of local perturbations in cellular networks.

2

Characteristics of Metabolic Networks

The availability of the complete genomes allow many new types of experiments and analysis. The new approaches and questions that the genomes make possible are usually referred to as functional genomics. The task is to define the function of a gene (or its protein) in the life processes of the organism, where function refers to the role it plays in a larger context. The most obvious example is the metabolism of an organism, the basic chemical system that generates essential components such as amino acids, sugars and lipids, and the energy required to synthesize them and to use them in creating proteins and cellular structures. This system of connected chemical reactions is a metabolic network.

Below is an example of such a network. This is the subsystem for synthesis of the amino-acids valine, leucine and isoleucine. This image was taken from the KEGG database. Signalling networks are another kind of network (or pathway) that are of central importance in biology. If the regulation of genes are involved, one often talks about genetic networks. Although the processes are different, and the types of proteins and other molecules are different, the overall picture is fairly similar: There are many components which interact in a connected network.

Typically, metabolic and signalling networks are viewed as different entities. In metabolic networks the flow of mass and energy is the essential purpose of the machinery. In signalling networks the purpose is the regulation of other processes, and the use of energy and mass flow is a requirement, but not really the point. However, there is an essential component of regulation also in metabolic networks: The enzymes are regulated through interactions with substrates and products so that the appropriate conditions in the cell are upheld.

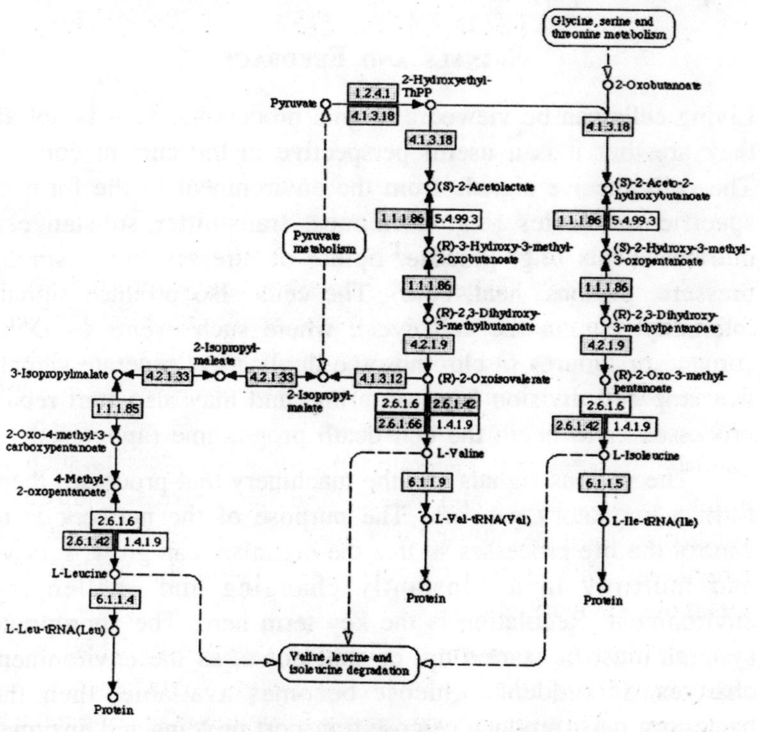

Valine, Leucine and Isoleucine biosynthesis

An important emerging field in bioinformatics is to understand metabolic and signalling networks in term of their function in the organism and in relation to the data we already have. This requires combining information from a large number of sources:

classical biochemistry, genomics, functional genomics (e.g. microarray experiments), network analysis, life process descriptions, and simulation. A theory of the cell must combine the descriptions of the structures in it (genome, proteome, subcellular structures, etc) with a theoretical and computational description of the dynamics of the life processes. One of the most important future challenges for bioinformatics is how to make all this information comprehensible in biological terms. This is necessary in order facilitate the use of the information for predictive purposes.

Signals and Feedback

Living cells can be viewed as signal processors. That is not all they are, but it is a useful perspective in the current context. The cells receive signals from the environment in the form of specific molecules (e.g. hormones, transmitter substances), nutrient levels (e.g. glucose, lipids) or stresses (e.g. osmotic pressure, poisons, heat, cold). The cells also produce signals internally, e.g. in the cell cycle, where such events as DNA damage or failures in chromosome duplication generate signals that stop cell division from occuring, and may also start repair processes or to begin the cell death programme (apoptosis).

The various signals and the machinery that processes them form a regulatory network. The purpose of the network is to control the life processes so that the organism can grow, survive and multiply in a constantly changing and challenging environment. Regulation is the key term here. The appropriate systems must be started up, or shut down, as the environment changes. If suddenly glucose becomes available, then the bacterium must produce glucose transport proteins and enzymes for the glucose metabolism.

If the glucose level is then reduced, the bacterium needs to reduce the production of these proteins since otherwise unnecessary energy is wasted on synthesizing the proteins. An important concept in this connection is homeostasis, the tendency of living systems to maintain a steady level of important metabolites or other components in the face of varying external circumstances. The most obvious example is the constant body temperature in many animals. An essential mechanism in regulation is feedback. The result of a process feeds back signals to control the process itself. This makes the process self-monitoring to some degree.

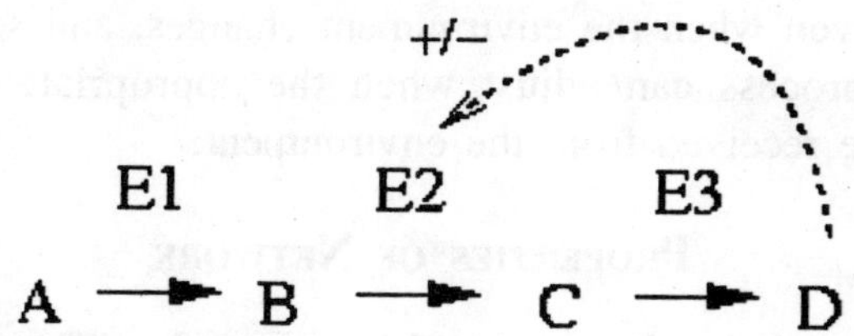

One distinguishes between negative and positive feedback. Negative feedback is a system where the product or resulting signal down-regulates the process. This may be useful to keep the level of a substance at a steady state. This is very common in the metabolism, since the organism only wants to produce as much as is needed of some particular compound, and no more. For example, many synthesis pathways for compounds such as amino acids are regulated by negative feedback, where the final product of the pathway influences some early step in the pathway negatively by inhibiting an enzyme. In the figure, compound D binds to enzyme E2, thereby inhibiting it, and reducing the further production of itself.

Positive feedback is a signalling loop where the product reinforces the process. This is found in systems which have to maintain one chosen state among several alternatives, e.g. the specific state of a differentiated cell, or to commit a cell towards a chosen development path. The starting point is an undecided state, which by some process (influence from the outside, or purely stochastic events) starts to tilt over in one particular direction. This small, initial bias is then reinforced by the regulatory system, so that a final, fixed, stable state is reached.

A special, very important class of regulatory networks generate cyclic processes. These are processes that never reach a steady state, but move between stages or states in an ordered fashion. There are many important examples of cyclic processes: The cell cycle is the prime example, but one could also mention the circadian clock (the changes following the time of day) in many plants and animals. These systems require complex networks having a dynamically cyclic behaviour that must both be robust and sensitive: Robust in the sense that it must continue to work even when the environment changes, and sensitive so that the process can adjust when the appropriate corrective signals are received from the environment.

Properties of Network

Metabolic and signalling networks are complex: There are many nodes (proteins, particles, molecules) and many connections (interactions). This is obvious from nearly any illustration of a network. And even if one can define sub-networks that can be meaningfully described in relative isolation (such as the citric acid cycle, or the ras p21 signalling network), there are always connections from it to other networks. The overall structure of a network can be described by several different parameters. For example, the average number of connections a node has in a network, or the probability that a node has a given number of connections.

Theoretical work has shown that different models for how a network has been created will give different values for these parameters. One can also plot the relative frequency of each node as a function of the number of connections of the nodes. Both axes in the plot uses logarithmic scales. Using data from real networks or from simulations of different types of randomised networks, one obtains log-log plots that turn out to have characterisic shapes depending on how the properties of the networks and how they were generated. Examples are shown in the figures below.

The classical random network theory by Erdös and Renyi states that given a set of nodes, the connections are made randomly between the nodes. This gives a network where most nodes have the same number of connections. The log-log frequency plot for this kind of network is shown below. Recent research has shown that this model does not fit the structure found in several important networks, such as the Internet, World-Wide Web, power grids, or social networks. Instead, these complex networks are better described by a so-called scale-free, hierarchical model where most nodes have only a few connections, but a few nodes (called hubs) have a very large number of connections.

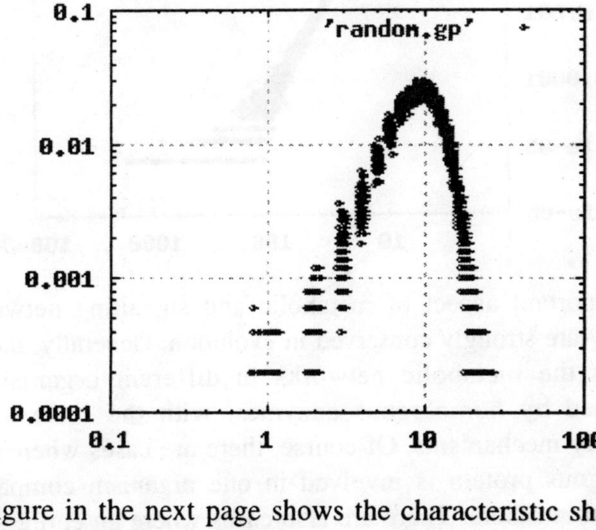

The figure in the next page shows the characteristic shape of a scale-free, hierarchical network in a log-log plot. This particular plot was produced by simulating a growing network, where each new node is connected randomly to existing nodes, but with a higher preference for already highly connected nodes. Robustness is another important property of metabolic and signalling networks. This is the ability of the network to produce essentially the same behaviour even when the various parameters controlling its components vary within considerable ranges.

For instance, recent work indicates the segment polarity network in the Drosophila embryo can function satisfactorily with a surprisingly large number of randomly chosen parameter sets. The parameters do not have to be carefully tuned or optimised. This makes biological sense: A signalling system should be tolerant with respect to mutations or large environmental changes.

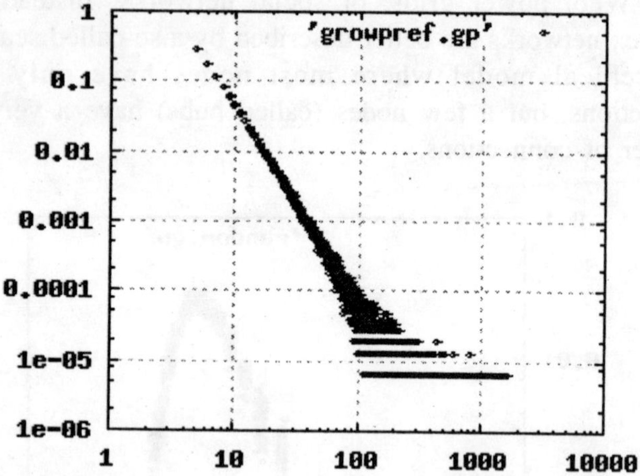

One important aspect of metabolic and signalling networks is that they are strongly conserved in evolution. Generally, the same roles in the metabolic networks in different organisms are performed by homologous enzymes, with the same kind of regulatory mechanisms. Of course, there are cases where a non-homologous protein is involved in one organism compared to the norm in others. There are also cases where an entire section of a network is missing in an organism.

Citric Acid Cycle

The citric acid cycle, also called the Krebs cycle, or TCA (tricarboxylic acid) cycle, is a central part of the respiratory energy metabolism of many organisms. A number of enzymes act on substrates in a cycle, where an acetyl group from acetyl-coenzyme A is covalently added to oxaloacetate and then converted into two molecules of CO_2 and reducing equivalents. In one of the first experiments using microarrays for monitoring

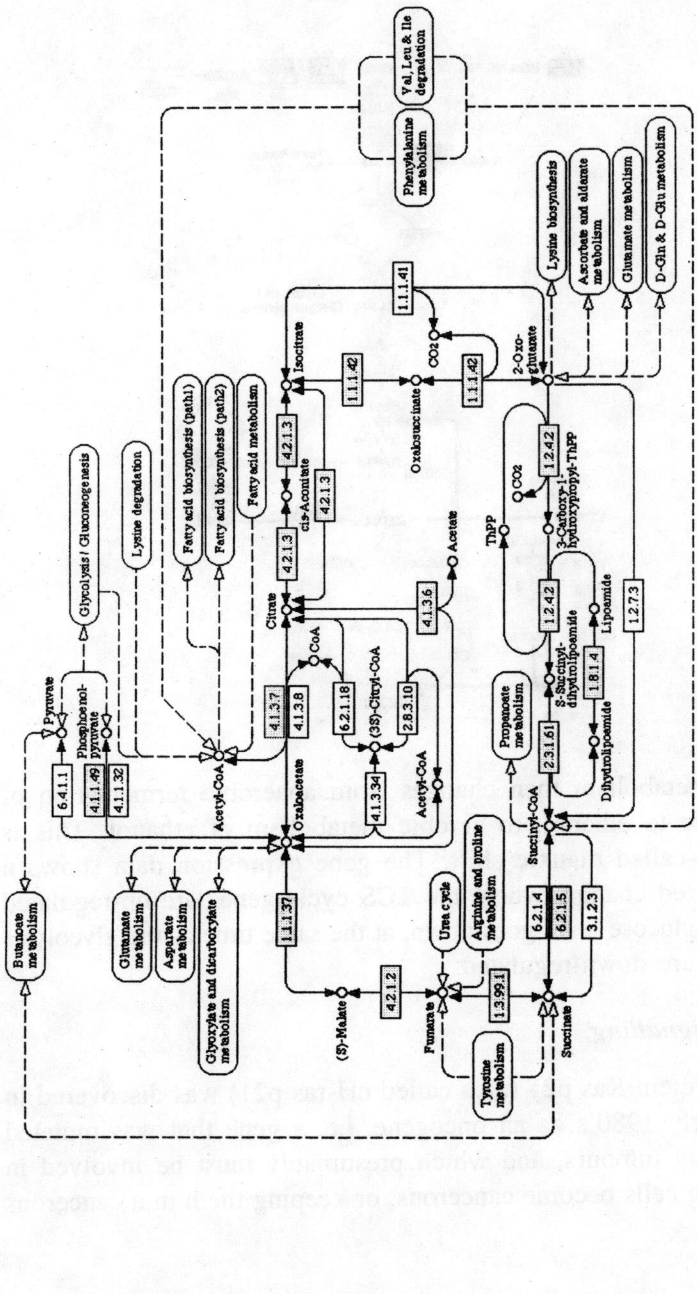

Citrate Cycle

gene expression changes, a yeast culture was allowed to grow in glucose-rich medium until the glucose ran out.

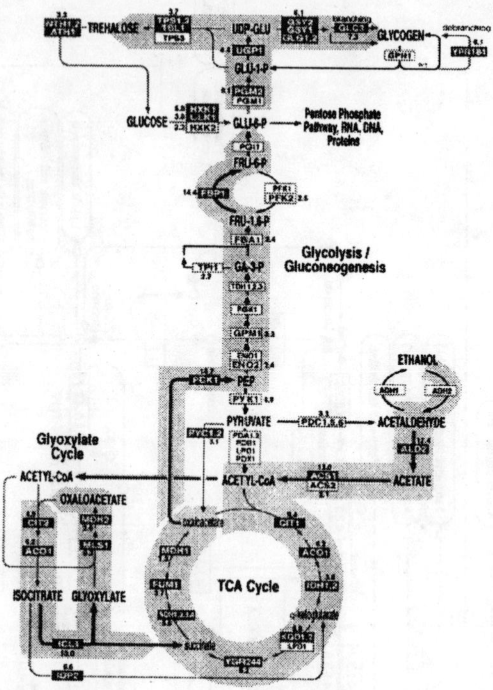

The metabolism then changes from anaerobic fermentation of glucose to ethanol, to aerobic metabolism of ethanol. This is the so-called diauxic shift. The gene expression data shows a concerted change, where the TCS cycle genes are up-regulated as the glucose level goes down, at the same time as the glycolysis genes are down-regulated.

Ras Signalling

The protein Ras p21 (also called cH-ras p21) was discovered in the early 1980's as an oncogene, i.e. a gene that was mutated in many tumours, and which presumably must be involved in making cells become cancerous, or keeping them in a cancerous state. An immense amount of research has gone into characterisation of these oncogenes, and one large group of them have turned out to be genes that participate in the regulation of cell growth and division. In the beginning of the 1990's, the growth regulation pathway which Ras p21 is a central part of was unravelled. It turns out that Ras p21 is part of a signalling cascade that also contains the proteins Shc, Grb2, SOS, GAP, Raf-1, MEK, ERK, and others.

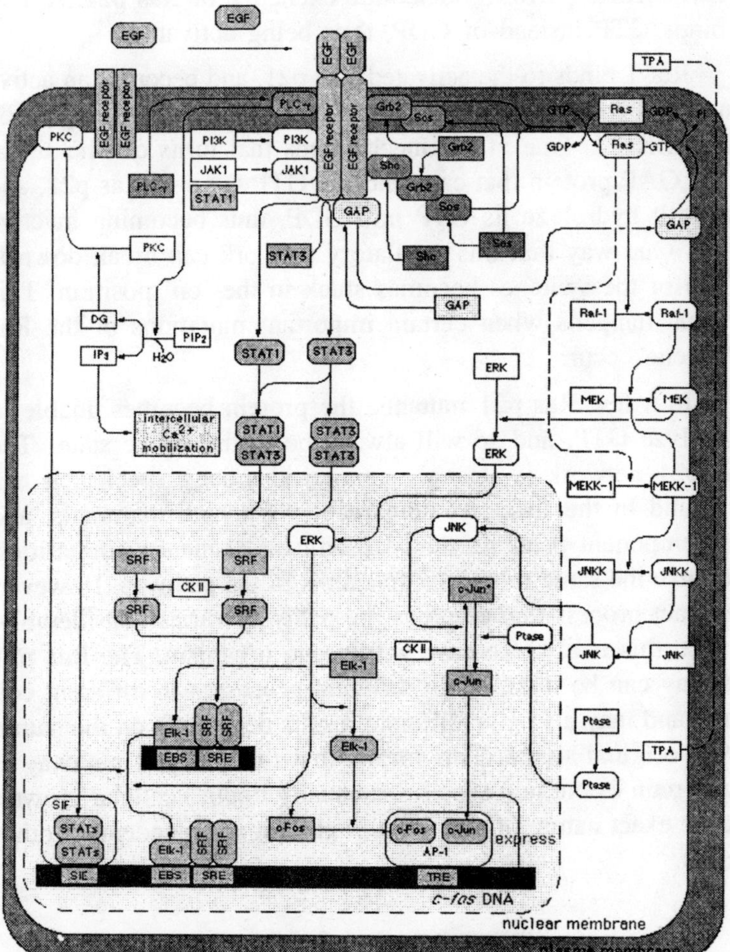

The purpose of the cascade is to transduce a signal from a growth-factor receptor in the cell membrane (there are several different ones) to the transcriptional regulation of genes in the nucleus that control cellular growth and division. In the image below (from the SPAD database), the receptor is the EGF receptor, which is a kinase that autophosphorylates itself upon binding of the EGF hormone. The Grb2 protein then binds the phosphorylated EGF receptor. This complex binds to SOS, which is activated to perform nucleotide exchange of Ras p21, so that it binds GTP instead of GDP, thus being activated.

Raf-1 binds to the activated Ras p21, and becomes an active kinase which turns on a phosphorylation cascade involving MEK and MEKK-1. One of the mechanisms that turns off the signal is the GAP protein that can bind the GTP form of Ras p21, and makes it hydrolyze its GTP into GDP, thus becoming inactive again. One way that this regulatory network can break down is if one of the switches becomes stuck in the 'on' position. This is what happens when certain important mutations of the Ras p21 gene occur.

In a few Ras p21 mutants, the protein becomes unable to hydrolyze GTP, and so will always be in the active state. The Ras p21 network is strongly conserved among species. It can be found in the fruit fly, nematode, yeast, and mammals, and the component proteins share significant sequence similarities, and have the same molecular functions in the pathway. However, the exact process that employs the pathway varies considerably. In fact, this is also the case within one organism. The Ras p21 pathway can be involved in different processes in different cell types, and may also give different results depending on the nature of the original signal. This means that the Ras p21 pathway is to a certain extent non-specific: it usually implies cellular growth, but the exact nature of that cellular growth is dependent on other factors.

3

Signalling Pathways

Neurobiology now has to face an unsettling concept: that hidden within "atomic" compartments and under all the ion channels, there exists a whole new universe of subcellular networks. These networks are the multitudes of interacting biochemical signalling pathways. They profoundly affect everything from the properties of single channels to the morphology of neurons and the wiring of the brain itself. Recent work on biochemical signalling reactions has emphasised the sheer complexity of these networks.

There are dozens of known major signalling pathways, each having at least five to ten enzyme isoforms, each of which communicates with a different subset of other messengers and pathways. As with neuronal networks themselves, these biochemical networks cannot be analysed in isolation. Neuronal signalling events from above, and nuclear and metabolic events from below, provide a barrage of signals that this network must process.

Metabolism, cytoskeleton formation, gene regulation, response to external inputs, differentiation, cell cycle, synaptic computation—all these operations are in the domain of signalling pathways. Consider a large industrial chemical complex. The control system for such a system is typically a huge computer, with hundreds of control points and sensors. A living cell is a chemical system whose complexity is orders of magnitude beyond anything man-made, and its control systems are correspondingly complex. Although a lot of the "software" is encoded in DNA, the role of "hardware" (more properly, "wetware") is carried out by biochemical pathways.

NOMENCLATURE

A signalling pathway, is any series of reactions that manipulate information of importance to a cell. This definition is rather broad, and might, for instance, be considered to include metabolic regulatory pathways that do not have much to do with the outside world. We are mostly concerned with pathways that do involve interaction with external events, particularly neuronal events. It is sometimes difficult to decide where one pathway ends and another begins. The term pathway to describe a group of reactions influencing a single signalling enzyme.

In this sense, then, the classical cyclic AMP pathway shown in Fig. 1 consists of at least four pathways: the receptor-ligand pathway, the G protein activation pathway, the adenylyl cyclase pathway, and the protein kinase A activation pathway. Another way of looking at it is to consider a pathway as a node at which information can converge, be processed, and sent on to multiple other nodes.

The term second messenger is typically used for a small signalling molecule (i.e., not a protein) that is produced indirectly through the action of some primary messenger such as a neurotransmitter. Examples are cAMP (cyclic adenosine monophosphate), DAG (diacylglycerol), IP_3 (inositol 1,4,5 triphosphate), AA (arachidonic acid), and, of course, Ca (calcium). Given the proliferation of interactions among signalling pathways, it is sometimes difficult to decide if a given molecule is a second, third, or nth messenger. We will apply the term to most small non-protein signalling molecules.

The term signal transduction will be applied in a restrictive manner to receptors that change or transduce one signalling modality (such as light) to another, such as biochemical signals.

In this sense a Ca channel is a sort of signal transduction mechanism, but the more specific term ion channel will be used in this case. Receptors will be used to describe proteins that detect neurotransmitters, usually at a synapse.

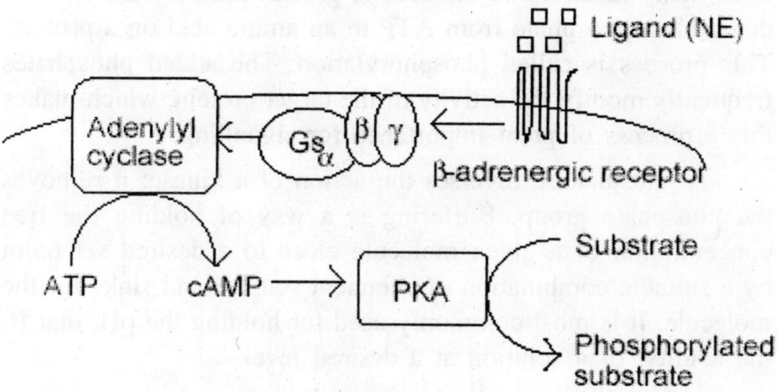

Figure 1. The cyclic AMP pathway. Thre are four successive pathways: (1) activation of the β-adrenergic receptor, (2) activation of the G protein G_s, (3) activation of adenylyl cyclase, and (4) activation of protein kinase A.

DNA (deoxyribonucleic acid) is the genetic material of the cell. It encodes information in the sequence of four chemical building blocks called bases. These are labeled A, T, C and G. A protein is a chain of amino acids (another building block). There are 20 amino acids. The sequence of amino acids is specified by DNA, and in turn determines the threedimensional structure and biochemical properties of the protein. Proteins frequently associate in pairs to form dimers, or triplets (trimers), or larger numbers (multimers).

An isoform of a protein is another protein, with a different but usually closely related sequence. Isoforms usually have similar enzymatic properties, but may be regulated differently. An enzyme is a protein with catalytic activity. It speeds up reactions. Enzymes usually exhibit great specificity in their substrates, the molecules they act upon, and also in the products they generate.

ATP (adenosine triphosphate) is the major source of chemical energy in the cell. The removal of one or two phosphates from ATP releases energy. This reaction is called hydrolysis. GTP (guanosine triphosphate) is a chemical cousin of ATP, and shares its energetic properties but is not involved in as many reactions in the cell. A protein kinase is an enzyme that adds a phosphate from ATP to an amino acid on a protein. This process is called phosphorylation. The added phosphates frequently modify the activity of the target protein, which makes this a process of great importance for signalling.

A phosphatase reverses the action of a kinase: it removes the phosphate group. Buffering is a way of holding the free concentration of a given molecule close to a desired set point by a suitable combination of chemical sources and sinks for the molecule. It is most commonly used for holding the pH, that is, the acidity, of a solution at a desired level.

Common Signalling Pathways

Regrettably, this section looks rather like an alphabet soup. It can be skimmed through on a first reading, but it is actually the barest of introductions to the topic. Anyone contemplating research simulations on signalling will need to go into much greater breadth and depth than this incomplete list. You should assume that there are tens of known examples in each category, each of which has tens of known isoforms. You should also be aware that the field is in an exponential growth phase: new isoforms are reported almost daily, and entirely new pathways and signalling mechanisms turn up every few months.

Receptors

These collect information from outside the cell, and couple it into intracellular pathways. There are two main subclasses of receptors. Ligand-gated ion channels are directly gated by receptors such as the familiar NMDA receptor, and pass signalling ions such as calcium into the cell when suitably stimulated by the presence of a ligand. G protein coupled receptors, such as the beta-adrenergic receptor, belong to a second category that gates channels indirectly. These receptors activate G proteins when stimulated by ligands. The G protein may then either modulate channel activity by binding directly to the channel, or regulate the activity of an enzyme involved in a second messenger pathway that modulates channel activity.

G Protein Pathways

The term G protein refers to proteins that bind guanine nucleotides, such as GTP. There are at least four different major G protein families, Gs, Gi, Go, and Gq. Gt (transducin), in the visual receptor pathway, is a G protein in the Gi family that is stimulated by light. As usual, each family has five to ten known isoforms. These trimeric, membrane-associated signalling proteins contain a, b, and g subunits. The b and g subunits have their own sets of isoforms, so there are plenty of permutations. Only a small fraction of these are believed to occur in nature. Activated a wanders around on its own, activating further pathways, but the bg complex remains dimerised. The prototypical G protein reaction is shown in Fig. 2.

The main steps in the cycle are as follows. First, the ligand activates the G protein coupled receptor. The activated receptor promotes the replacement of GDP by GTP, following which, the active Ga-GTP splits off from the Gbg dimer. The active Ga binds to an effector enzyme such as adenylyl cyclase, and turns it on. In parallel, Gbg binds to its own target enzymes or channels. After a while (a few seconds to minutes), the slow GTPase activity of the Ga hydrolyzes the GTP to GDP.

Now the pieces reassemble: Ga-GDP and the Gbg associate, and the inactive heterotrimer can bind to another receptor to repeat the cycle. There are two major features to note here. First, the G protein loop amplifies signals. A receptor can catalyse the activation of numerous G proteins while the ligand is bound. Each Ga remains active until the GTP is hydrolysed, and this

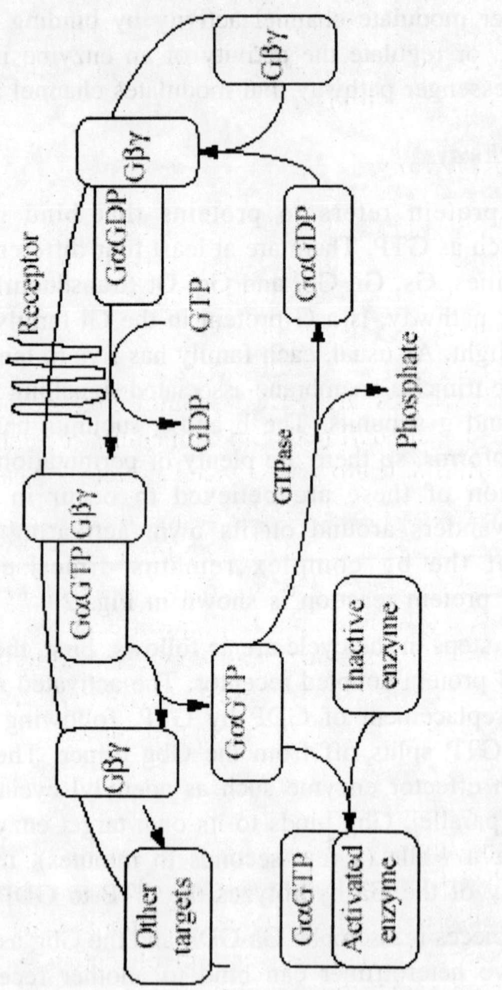

Figure 2. G protein signalling cycle.

takes a minute or so. During this time, the activated effector enzyme can process a large amount of substrate. Second, there are two signalling outputs from a G protein: the Ga and the Gbg. Each can potentially influence several signalling pathways.

There is also a large subfamily of "small G proteins," which lack bg subunits. These include Ras, Rab, and many others. They have their own family of regulatory proteins that affect the rate of GTP turnover.

Protein Kinase Pathways

Protein kinases add phosphate groups to specific sites on other proteins, and thereby affect their activity. There are dozens of kinases known, which fall into two main classes: serine/threonine kinases and tyrosine kinases. These classes indicate the amino acids that are phosphorylated by the kinase. The major serine/threonine kinases are PKC (protein kinase C), PKA (protein kinase A), CaM-KII (calcium-calmodulin regulated type II kinase), and MAPK (mitogen-activated protein kinase).

Receptor tyrosine kinases (RTKs) are an important class of tyrosine kinase. Beyond the specificity for a particular amino acid substrate, each kinase has its own range of sequence preferences, which can be exquisitely specific or very broad. Kinase activity is regulated in many ways, including second messengers such as Ca, DAG or cAMP, phosphorylation, membrane association, or even direct ligand binding.

Phosphatases

These reverse the activity of kinases: they remove phosphate groups. Like kinases, they have their own classes, isoforms, substrate specificities and so on. Major phosphatases are PP-2A (protein phosphatase 2 A), PP-1 (protein phosphatase 1) and calcineurin. It used to be thought that their only role was to balance out kinase function, but it is now clear that they are in turn regulated in a wide variety of ways and contribute actively to signalling.

Phospholipases

These take phospholipids, which are a major constituent of the cell membrane, and turn them into active signalling molecules. Examples include PLA_2 (phospholipase A_2), PLC-β (phospholipase C-β), and phospholipase D. PLC-β is particularly interesting as it produces two signals: diacylglycerol (DAG) and IP_3 (inositol triphosphate). PLA_2 produces AA, which is a membrane-permeable signal and may be involved in retrograde signalling. It should be pointed out that many phospholipids themselves are important in signalling.

Cyclases

The main cyclases are AC and GC (adenylyl and guanylyl cyclases), which produce the cyclic nucleotides cAMP and cGMP from ATP and GTP, respectively. There are at least ten ACs known, each of which has its own regulatory preferences. All ACs are activated by $G_{s\alpha}$; most are inhibited by $G_{i\alpha}$, and there is regulation by bg, phosphorylation, and so on, in various combinations.

Phosphodiesterases

These degrade cAMP and cGMP, and thereby turn off signals from cyclases. Regulatory mechanisms include CaM-dependence and phosphorylation.

Calcium

Although calcium is just a simple ion, and not really a pathway, it is arguably the most important single signalling molecule. The regulation of Ca alone involves ligand and voltagegated channels, pumps, intracellular stores, buffers, diffusive microenvironments and more. It is involved in the regulation of almost all kinds of pathways, and there are whole families of proteins such as CaM (calmodulin), whose only role is to detect Ca levels as part of the regulation of other proteins.

MODELLING SIGNALLING PATHWAYS

Theory

Chemical rate theory is an old, well-established subject. A lot of it has to do with analytical derivation of results that we, in our modern-day decadence, leave to computers. For this purposes we have to understand is a single rate equation:

$$A + B \underset{k_b}{\overset{k_f}{\rightleftharpoons}} C + D. \tag{1}$$

By definition, this is described by a differential equation of the form

$$d[A]/dt = k_b[C][D] - k_f[A][B], \tag{2}$$

where the square brackets indicate the concentration of the specified molecule. The mass conservation laws constrain the remaining concentrations, once the starting concentrations and the current value of [A] are known. You have already encountered very similar rate equtions when studying the Hodgkin-Huxley model of activation of sodium channels. Enzyme catalysis is a very important part of biochemical reactions. One of the simplest and best descriptions of enzyme kinetics is due to Michaelis and Menten. Way back in 1913, they described catalysis as a two-step process, where the enzyme and substrate first reversibly combine to form a complex. This complex can then can go forward to form the product in an essentially irreversible step:

$$\text{Enz} + \text{Sub} \underset{k_2}{\overset{k_1}{\rightleftharpoons}} \text{Enz.Sub} \xrightarrow{k_3} \text{Enz} + \text{Prd}. \tag{3}$$

Although we need three rate constants here, to specify only two parameters which do an excellent job of representing enzyme properties. These are:

$$K_m = (k_2 + k_3)/k_1 \tag{4}$$

and

$$V_{max} = k_3. \tag{5}$$

K_m is the Michaelis-Menten constant for the enzyme. It is the concentration of substrate at which the rate of generation of product is half maximal. V_{max} is the maximum velocity of the enzyme, i.e., the rate of formation of product when there are saturating amounts of substrate present. All of the enzyme will then be complexed with the substrate, so V_{max} is just k_3.

Signalling pathways are typically represented as neat little black boxes connected with arrows. The first step in constructing a model of such a pathway is to open up the black box and recoil in horror at the seething mass of interactions inside. These interactions, which are frequently ill-defined in mechanistic terms, must be framed in terms of individual chemical reactions. In GENESIS, these are binding/reversible reactions, and Michaelis-Menten enzyme reactions. Complete mechanistic details for signalling pathways are available in only a few cases. G protein signalling, for example, has been worked out in considerable detail (Fig. 2).

For example, several important kinases including PKC and PKA are known to incorporate an enzyme site and a pseudosubstrate region, which binds to and inactivates the enzyme site. As a modeler, one may have to make compromises on several fronts:

— If one is lucky, the mechanistic data may actually go beyond what you need or can handle. PKA regulation is a case where there is such an overabundance of information. Judicious simplification is usually not difficult in such cases.

— At the other extreme, mechanistic details may be so sparse that you have to develop an "empirical" model, which uses the simplest possible mechanisms that fit the observed data.

This approach works surprisingly well, especially if there are plenty of data to constrain the system. A common situation is where there are several known activators for an enzyme, each of which is described by a different concentration-effect curve.

The dumbest approach is then to treat each activation process as a separate reaction, resulting in an activated enzyme with different rates. This has been done, for example, for the activation of PLA_2. One can often go a step further and rule out certain possible mechanisms on the basis of such raw concentration-effect curves.

The general, and most common case, is that some of the crucial mechanistic details are known, and the rest are a bit fuzzy. One can often fall back on mechanisms for similar pathways to fill in the gaps. Otherwise, one just has to plug the holes with the simplest mechanisms that will work.

As an example of figuring out mechanistic details, consider protein kinase C. It is known to be regulated by a pseudosubstrate domain that normally binds to and blocks the kinase site. Activators function by releasing this blockage, in assorted ways. This crucial regulatory information enables us to predict that most activators will expose an enzyme site of identical activity. This implies that the strength of the activator is likely to depend on its efficacy in releasing the block, rather than on special properties of the enzyme-activator complex. So, the active enzyme can be represented as a single pool, to which different activators contribute an amount determined by their respective efficacy. To round off the picture, let us consider activation by Ca, DAG, and AA.

This lets us draw up the first skeleton of the activation process (Fig. 3). We just need to sum up the individual contributions into a total final activity, as we assume that the exposed enzyme site is identical in all cases. We can further elaborate on the activation process by noting that each of the activators works synergistically with the others.

One way of modelling this might be to add further reactions where the two activators combine, as illustrated by the Ca-AA synergistic pathway in dotted lines. As it turns out, we can considerably refine the activation mechanisms because the concentration-effect curves require even more complex interactions for the model to fit well.

Reaction Rate Constants

When a signalling pathway is modeled in terms of biochemical reactions, the immediate parameters required are rate constants and initial concentrations. Unfortunately, biochemists rarely provide parameters with modelers in mind. The most common form for expressing biochemical data is the concentration-effect curve (Fig. 3B) which describes an effect, such as activation, rate of production, or some similar experimental quantity, in terms of the concentration of an activating agent.

Mechanistically, the link between the concentration and the effect is quite likely to involve numerous intermediate steps. By the time you get to the point where you want to fill in rates, we assume that you have a specific mechanistic model in mind, so you should have an idea of what these steps are. Obviously, you want to start with the simplest curve involving the fewest unknowns. If one must express a concentration-effect curve as a single number (losing information in the process) it is the concentration at half-maximum, K_d. Assuming that your reaction is first-order, $K_d / k_b / k_f$. If in addition one has information about the time-course τ of the reaction, one can completely define kb and k_f. This procedure is fine in theory. In practice, if you do not actually run your simulation and compare the concentration-effect plot point-by-point, you will throw away a lot of valuable data and probably lose the chance to catch serious mistakes in

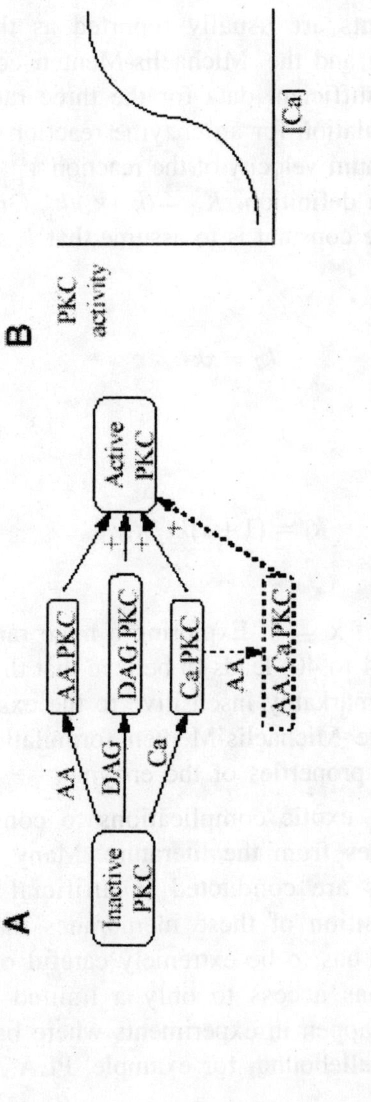

Figure 3. *(A) Skeleton model of PKC activation. (B) Example concentration-effect curve.*

your assumptions.. Such sets of curves embody enormous amounts of information to help constrain rates, mechanisms, and interactions between different activators.

Enzyme Rate Constants

Enzyme rate constants are usually reported as the maximal enzymatic rate Vmax and the Michaelis-Menten constant Km. This is obviously insufficient data for the three rate constants in the standard formulation for an enzyme reaction. As already mentioned, the maximum velocity of the reaction $V_{max} = k_3$. From the Michaelis-Menten definition, $K_m = (k_3+k_2)/k_1$. One option to fill in the missing rate constant is to assume that k_3 and k_2 are in a fixed ratio:

$$k_2 = xk_3. \tag{6}$$

Then,

$$k_1 = (1+x)k_3/K_m. \tag{7}$$

We can use a value of x = 4. Exploring a huge range of such scale factors (from 0.4 to 40) leads to believe that the behaviour of most models is remarkably insensitive to the exact value of x. In other words, the Michaelis-Menten formulation captures most of the essential properties of the enzyme.

There are some exotic complications to consider when obtaining enzyme rates from the literature. Many membrane-bound measurements are conducted in artificial membrane systems. The composition of these membranes may strongly affect rates. One also has to be extremely careful of situations where the enzyme has access to only a limited amount of substrate. This may happen in experiments where both enzyme and substrate are micellebound, for example, PLA_2.

Initial Concentrations

Determining initial concentrations for signalling molecules is a task fraught with peril. In principle, one just has to look up purification stages for enzymes, and standard biochemical measurements for substrates and messengers. In practice, one needs to watch out for:

- — Tissue and species specificity
- — Partitioning between different fractions of tissue (membrane, cytosolic, particulate, etc.)
- — Purification methods, and yield and loss of activity during purification
- — Loss of activity (or sometimes even gain of activity!) during storage and many other complications. Given all these hazards, the more references for each concentration, the better.

Typically, one will need to provide only a few initial concentrations and just let the simulation run to steady-state to obtain the remainder. This is essential, as it is experimentally very difficult to obtain steady-state values for many of the reaction intermediates. If one has equilibrium values for some molecules, that is a bonus and an excellent cross-check.

Refining the Model

There is no final step to building a good model of any kind. In signalling models in particular, it takes many iterations to converge to a good representation of the system. All the preceding steps, (except hopefully the theory) will need to be revisited as the model evolves. An important part of this refinement includes repeated scans through the literature to hunt out alternative sources for each parameter, and cross-checks that become feasible as the model becomes more predictive.

Kinetics Models with GENESIS and Kinetikit

The modules for kinetic modelling are provided by the kinetics library, whereas Kinetikit (or kkit) is the user interface. Kinetikit is similar to Neurokit in the sense that it is a research tool rather than a tutorial programme. It makes much better use of the features provided by XODUS.

Kinetics Library

A GENESIS library contains the definitions of commands that will be accepted by GENESIS, as well as basic building blocks, called objects, which are used for the construction of simulations. The kinetics library defines additional commands and objects to extend the capabilities of GENESIS to simple models of biochemical signalling.

There is ample complexity in biological signalling, even when such vital complications as diffusion, compartmentalisation, and enzymatic scaffolding are not addressed. A simple exponential Euler scheme is used by the kinetics library. Fortunately, stiffness does not seem to be such a problem in the kinetic computations. Time steps of 2 msec are sufficient for most simulations of biochemical reaction systems. It is trivial to implement a crude variable time step scheme when known steep stimuli occur-just lower the time step.

About 100 *μsec* will usually give you sufficient stability and accuracy for such cases. A *ksolve* object (cousin to the *hsolve*) is being designed to use much faster implicit and variable time step techniques. Even the current methods work many times faster than real-time for all but the most complex simulations. All calculations are carried out on numbers of molecules, rather than concentrations.

This simplifies computations where molecules are exchanged between chemical compartments of different volumes. However, concentrations are calculated locally by each molecular pool, by dividing the number of molecules by a vol field of the pool object. The vol field is typically scaled to include Avogadro's number so that the concentration units are μM or some other familiar units.

The kinetics library assumes bulk quantities of all molecules. Markov models are specifically not simululated in the current kinetics library. Caution should therefore be employed when dealing with tiny compartments: it has been estimated that there are about ten Ca ions in a small dendritic spine. Mean rate theory may not apply in such situations.

The GENESIS objects provided by the kinetics library are the pool of molecules, the reac for simple reactions, and an enz which can be attached to a reactant pool to provide an enzyme site. A simple concchan object is provided for constructing concentration-driven channels within the kinetics library.

Kinetikit

The method of choice for developing kinetic simulations in GENESIS is to use Kinetikit, which is a graphical interface and simulation tool designed specifically to make it easy to build and manage complex kinetic simulations. To get you started quickly, we'll go through a demonstration simulation that illustrates many of its features. This demonstration involves a simple reaction where a molecule A reversibly converts into a molecule B:

$$A \underset{k_b}{\overset{k_f}{\rightleftharpoons}} B. \tag{8}$$

Step 0

Make sure you have a version of GENESIS that includes the kinetics library. If there is a startup message of the form "The kinetics library is copylefted under the LGPL, see kinetics/ COPYRIGHT." then it is there. If not, you will need to follow the installation steps detailed in the kinetics documentation. Once you have it installed, you are ready to go. After changing to the directory in which Kinetikit resides (usually Scripts/kinetikit), perform the following steps.

Step 1

Type:

```
genesis kkit
```

This will display a start-up window detailing the philosophy behind Kinetikit models, and also indicating the licensing terms. As soon as Kinetikit has loaded, this window will vanish and will be replaced by a screen similar to that in Fig. 4. The control window includes various menu options on the top.

The simulation run control buttons in bright traffic-light colors are just below, and dialogs for the simulation duration and current time below them. At the bottom is a row of strange icons. These are the building blocks for a kinetics simulation, and their role will shortly become clear. Below the control window is the Kinetikit edit window, where the reactions are set up. It will initially be blank. The graph windows are the only other windows currently visible.

Step 2

Use the left mouse button to click and drag the first of the icons, the *kpool* icon, into the edit window. Two things should happen: a *kpool* icon should appear in the edit window, and a new window should pop up with lots of dialogs for parameters. What you have just done is to create a new pool of molecules in the simulation, and to help you along, the parameter window has been displayed. Typically the first thing one does is to change the color and name of the icon to something that suits you.

Step 3

Familiarize yourself with the properties of the edit window. Move the *kpool* icon in the edit window around a little, using

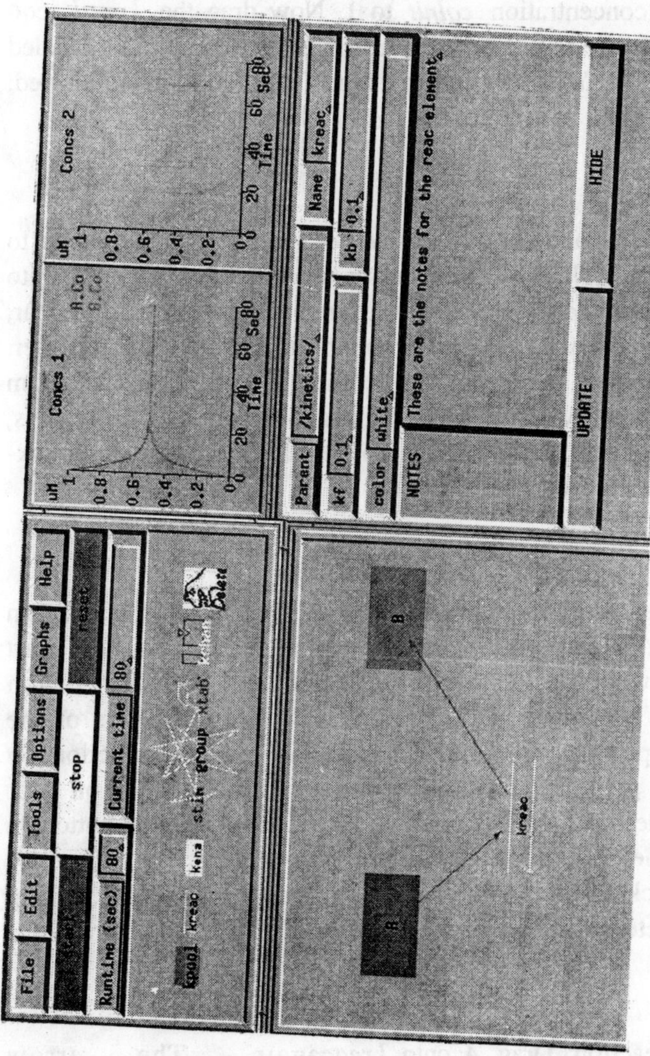

Figure 4. *Screen dump of Kinetikit*

click-and-drag with the mouse. The layout of the reactions has nothing to do with the numerical calculations, but a clean layout makes it much easier to figure out what is going on. Now double-click on the *kpool* icon. The parameter window reappears. Set the initial concentration *colnit* to 1. Now drag the *kpool* icon to the graph window. You have just made yourself a plot, labeled *A.Co*. The "*A.Co*" suffix indicates the field that is being plotted, in this case the concentration *Co* of element A.

Step 4

Drag in a *reac* and another *kpool* (let's call it *B*). Add the *B* to the graph as well. Now we can set up a reaction. Drag *A* onto *reac* and then *reac* onto *B*. Little green arrows should appear, linking them. If your reaction components are too close together, the little arrows don't fit, and so they vanish. You can zoom and pan the edit window using the angle-bracket and arrow keys, to see this effect. Also try rearranging the reaction using click-and-drag within the edit window.

Step 5

Run the reaction. Hit the *Start* button, and watch the reaction proceed. You can hit the *stop* button in mid-stream, and start up again without affecting the calculations. The *reset* button clears the simulation. Play around with the parameters of the pools (esspecially *CoInit, nInit* and *vol*), and the reaction (*kf* and *kb*) to get a feel for how it all works. Double-click on *A* or *B* while the simulation is running, or hit the *UPDATE* button in the parameter window, to monitor concentrations. Also try double-clicking on the graph axes and on the plot labels, to pop up parameter windows for specifying display parameters.

Step 6

Edit the reaction. Drag *A* onto *kreac* again. The arrow disappears. Repeat the drag, and it reappears. This is fine for simple editing, but what happens if you want to make it a second-order reaction 2A $\rightleftharpoons$ B? You will have to go to the *Options* menu and enable higher-order reactions to accomplish this feat. Now drag *kreac* onto Barney, the evil dinosaur icon in the control window. Chomp! Barney has deleted *kreac*. This works for everything in the edit window, and also for plots.

Step 7

Save the reaction. Click on the *File* menu button, fill in some notes, and enter your chosen filename, e.g., "*reacs. g*." It will be a script file, so you should have the usual "*g*." suffix.

```
genesis reacs.g
```

A Feedback Model

As a more complete example of the development of a kinetics simulation, we will use Kinetikit to come up with a feedback pathway that exhibits some interesting properties such as bistability. We then show how to run this model without the interface, and finally how to hook this up to other GENESIS components.

The feedback model consists of two enzymes, X and Y, each of which is activated by the product of the other enzyme. The activation process is second-order. The substrates are buffered: their concentration is fixed no matter how much is used up. The products are degraded in a simple first-order reaction to avoid runaway feedback. The reactions are represented in Fig. 5. The figure has almost a one-to-one relationship with the *Kinetikit* implementation, so it should be easy to implement. The saved model is available in the *kinetikit/* examples directory as *feedback.g*.

Following are the steps up the feedback model

1. All enzyme and reaction rates are 1.0, except for the degradation reactions, which have a k f of 0.16 and a kb of

0. Note that these are not the default values, so you will have to explicitly set them.

2. All initial concentrations are 0.0, except the X and Y substrates, which are buffered to an initial value of 1, and the inactive forms of X and Y, which are at an initial value of 1 but are not buffered.

3. Two molecules of product bind to one molecule of inactive enzyme to activate it. Look up the *Options* menu to see how to do this sort of second- and higher-order reaction. All other reactions are first-order.

4. The arrows with a double bend represent enzyme reactions.

5. The default time step of 0.01 is good for running the model.

6. Plot out the concentrations of X and Y to monitor the simulation. You can also double-click X or Y at any time to read out the concentration from the *Co* dialog.

7. Do not use spaces in the names of the components in the model. It will confuse the simulator when it tries to restore simulations from a file.

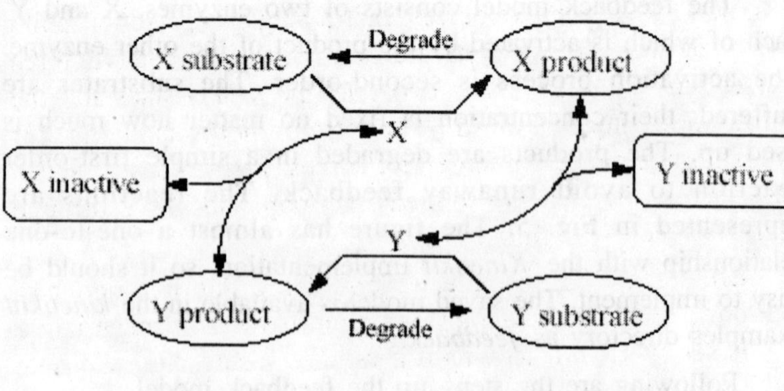

Figure 5. Reaction mechanism for feedback model

Bistability

As a preliminary check on your model, just run it for, say, 100 seconds. X and Y should remain steady at a concentration of 0.0. The steady-state value you have now reached is the lower bistable point. To convince yourself that there is an upper steady-state mode, dump some X product into the system by setting its *Co* to 2, using the dialog box that will appear when you click on *x-prd.* Watch the system climb up to the upper activated state. Let it run a while longer, and convince yourself that the system has now stabilised at a new level. Note this level. If you are ambitious, you can now figure out how to push the system back and forth between the two stable levels.

Concentration-Effect Curves

There is a simpler, experimentally more feasible, and theoretically informative way of characterising a biological feedback system of this kind. This technique works by "breaking" the loop at some point, and plotting concentration-effect curves of X vs. Y and vice versa. When the two curves are plotted on the same axes, so that the X vs. Y curve now has been flipped around, their intersection points define the behaviour of the system.

You can convince yourself that a single intersection point always defines the steady-state, and that if you have three intersection points the outer two must be steady-state and the middle one the threshold. These two methods using for generating the concentration-effect curves. In each case, we can buffer one of the enzymes to break the loop, and step it through a series of concentration values.

Brute force The simple way of doing this is to double-click on *x*, and flip the *buffer* toggle on. Now the concentration of X is set by *CoInit*. Run until it reaches steadystate, and note the concentration [Y]. Increase X, and repeat until Y saturates. Now go through the whole process again, with Y buffered and X free. Hmm . . . The numbers look familiar. Could you have predicted this?

Using xtab If you have to generate the concentration-effect curve more than once, say, for different parameters, you will probably wish to automate the whole process. The *xtab* object is good for this. This is an extended object which is created by Kinetikit. Drag in an *xtab* icon, and connect it up to *x*. Set up the *xtab* waveform to *x* a "staircase" of levels, with appropriate times for each step. You may need to refer to the Kinetikit documentation to help you with this. Activate the *xtab*, and run the simulation again. Your concentration-effect curve will be generated without further effort on your part.

4

Networks in Genomics and Proteomics

Complex multicellular organisms contain large genomes in which each structural gene is associated with at least one regulatory element and each regulatory element integrates the activity of at least two other genes. The nature of such regulation started to be understood from the analysis of small prokaryotic regulation subsystems and the current picture indicates that the webs that shape cellular behaviour are very complex.

Actually, integration of extracellular signals often involves the crosstalk between signal cascades that has been suggested to share some common traits with neural networks. In a related context, detailed analyses of subsets of interacting genes reveal that cell biology is highly modular. Here "modules" are made up of many species of interacting molecules and the functional relevance of these subnets is highlighted by the observation that they are conserved through evolution.

In many cases, proteins composed by multiple subunits behave as switch-like elements that can flip, for example, from an active to an inactive state and back. The switching behaviour of these complexes, together with the underlying information processing that takes place at the network level, allows for a computational description of intracellular signalling.

The analysis of mutational robustness in different organisms revealed an extraordinary level of homeostasis: in many cases the total suppression of a given gene in a given organism leads to a small phenotypic effect or even to no effect at all. Following the analogy with engineered systems, the immediate explanation for such robustness would come from the presence of a high degree of redundancy. Under mutation, additional copies of a given gene might compensate the failure of the other copy. However, the analysis of redundancy in genome data indicates that redundant genes are rapidly lost and that redundancy is not the leading mechanism responsible for mutational robustness.

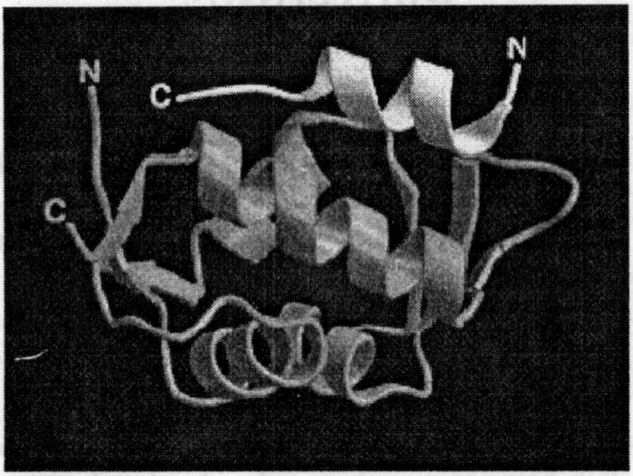

Figure 1. The domain of molecular interaction among p53 and MDM2 is shown in this 3D reconstruction. MDM2 (here in cyan) binds a specific domain of p53 in a region (shown in white) important for the interaction of p53 with components of the transcription machinery.

The origins of robustness against mutations is particularly well highlighted by the analysis of genome-wide scale data of the budding yeast Saccharomyces cerevisiae. Major cause of robustness comes from the interactions among unrelated genes. This mechanism would be illustrated by the following example: given metabolic network, completely unrelated enzymes can catalyse different reactions but contribute to a pathway whose goal is to sustain an optimal flux of metabolites. Under these conditions, mutations in genes encoding those enzymes will have little or mild effects.

Additionally, it is interesting to see that many examples of experimental biotechnology manipulations involving the tinkering of one or two genes fail to reach the expected goals: very often, counterintuitive outcomes are obtained. On the other hand, mutations involving some key genes can have very important consequences. This is the case, in particular, of the p53 tumour suppressor gene, Figure 1, which is known to play a critical role in genome stability and integrates many different signals related to cell-cycle or apoptosis (cell death). This and other tumour-suppressor genes prevent cell proliferation (thus keeping cell numbers under control) but can also promote apoptosis.

The example of p53 is particularly important because it is mutated or there is a functional defect in the p53 pathway in approximately half of human cancers. The p53 network (partially shown in Figure 2) is quite well-known in mammals and involves genes that control, for example, apoptosis, the development of blood vessels, or cell differentiation. The core of this net is defined by the feedback loop existing between p53 and its negative regulator, the MDM2 oncoprotein. In invertebrates (such as Drosophila) homologues of p53 are known to be active throughout early development. The fact that many mutations have little or no effect seems to be consistent with the presence of genes that either cannot propagate their failure or whose function can be replaced by other parts of the net.

The presence of some genes that integrate multiple signals and can trigger widespread changes under their failure shows that the underlying network includes some highly-connected hubs. It seems to be a compromise between integration and homeostasis that should be observable when looking at the map of interactions within the cellular net. Although a complete description of cellular networks would require the explicit consideration of dynamics, topological approaches-in which only the static architecture of the net is considered-are often successful in providing insight into biological complexity.

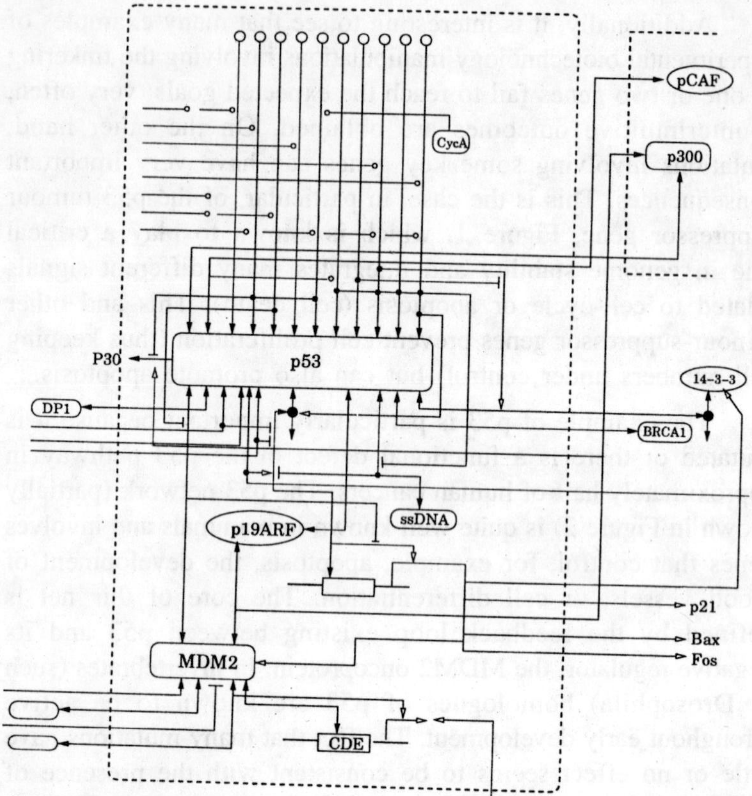

Figure 2. Schematic architecture of the p53 network. The p53 node integrates information from very different parts of the system. Only part of the cell circuitry is shown here.

For example, this is the case of some models of ecological networks: in spite that populations fluctuate in time and changes in biomass or productivity take place at different scales, some of the fundamental regularities exhibited by food webs can be fairly well explained by means of static approaches. Besides, the comparison of a wide range of complex networks reveals that strong regularities are shared by them, in spite that their underlying components, the nature of their interactions, and their time scales are very different.

MATHEMATICAL APPROXIMATIONS OF LARGE-SCALE MODELS OF CELLULAR NETS

The molecular basis of genetic control in cells, particularly in eukaryotic cells is one of the most basic active areas of molecular cell biology. Of particular interest is the understanding of the regulation mechanisms involved in the development of multicellular organisms. In most well-known case studies, it has been shown that regulation among the genes that control early development (Figure 3) takes place at the transcription level.

The web of interactions can be very complex, and an example of a sub-web of the genetic net associated to Drosophila early development is shown in Figure 3(b). Mutations in genes associated to early stages of development have typically a strong effect and sometimes, as it occurs with the so-called homeotic genes, they result in important morphological changes. Models of gene regulation have a long history in theoretical biology. The discovery of the mechanisms of transcription regulation in the Lac operon of E. coli was followed by the formulation of some simple mathematical models.

Inspired in early models of neural networks, a standard formulation of gene regulation can be introduced by means of a dynamical system:

$$\frac{dg_i(t)}{dt} = \Phi^i_\mu[\mathbf{g}] - \gamma_i g_i, \qquad i = 1, ..., n, \tag{1}$$

where a set of n different genes is defined. Here $g = (g_1 ..., g_n)$ gives the activity state of each gene. Degradation is introduced by the last term $\gamma_i g_i$. The function $\Phi^i_\mu(g)$ introduce the nature and extent of the interactions among components. An example of such type of model is:

$$\frac{dg_i(t)}{dt} = \Phi^i_\mu \left(\sum_{j=1}^{n} W_{ij} g_j(t) - \theta_i \right) - \gamma_i g_i(t), \tag{2}$$

where $\Phi^i_\mu(x)$ is a sigmoidal function of the local field $h_i = \Sigma_J W_{ij} g_j$, θ_i on is a threshold, and the weights W_{ij} give the sign and strength of the gene-gene interactions. Usually the set $W = \{W_{ij}\}$. As is generated from a given distribution $\rho(W)$ that is assumed to be symmetric and with zero mean. This type of net can display a huge variety of dynamical patterns, including oscillations and chaos. But the really interesting behaviour comes from the statistical properties derived from the presence of phase transitions when the connectivity is tuned.

Some attempts of building large-scale models of cellular nets based on near-realistic descriptions have failed to reproduce the whole spectrum of dynamical patterns displayed by even simple controlled systems.

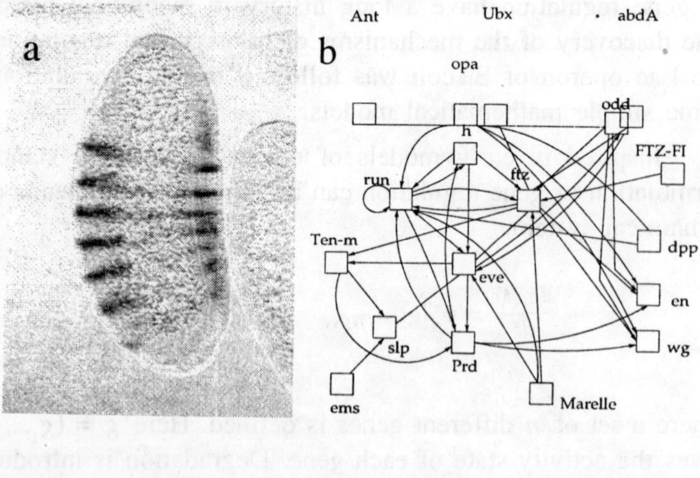

Figure 3. (a) Spatial pattern of activity of a given gene involved in Drosophila *development (the so-called fushi tarazu gene (FTZ).*

On the other hand, some key questions can find powerful answers in the generic properties exhibited by simple representations of real nets. As an example, a striking feature of multicellular diversity is the surprisingly small repertoire of cell types, given the potentially astronomic diversity of cell states that would be obtained from the combinatorics of gene states. Assuming that the number of genes in a multicellular organism is $N \approx 10^4$, 2^N different possible states are available. Yet, if cell types are considered as indicators of gene expression states, only 200- 3000 states are actually realised.

In figure 3, the darker areas correspond to higher levels of activity of FTZ, indicating what cells are expressing it. Cell-to-cell interactions generate this set of stripes with a characteristic length. In (b) an example of a real gene network is shown. It includes some part (i.e. a directed subgraph) of the genetic net involved in the development of *Drosophila* fly. The names of the genes involved are indicated, such as FTZ = fushi tarazu. Only the connections are shown, not their sign.

Two-gene Networks

The minimal number of genes needed in order to obtain a rich spectrum of behavioural patterns is given by two elements in interaction, although single-gene models with the appropriate nonlinearities can also display complex dynamic behaviour. Two-gene models allow to understand particularly important problems, such as the dynamics of virus-cell interactions in bacteria. An example is the following two-gene system with no self-interaction, described by the equations:

$$\frac{dg_1}{dt} = \delta \frac{W_{21}g_2}{1 + W_{21}g_2} - g_1 \tag{3}$$

$$\frac{dg_2}{dt} = \delta \frac{W_{12}g_1}{1 + W_{12}g_1} - g_2. \tag{4}$$

The fixed points are easily found; together with the trivial fixed point, $P_0^* = (0, 0)$ we get a second nontrivial point $P_1^* = (g_1^*, g_2^*)$ given by:

$$g_1^* = \frac{\alpha}{W_{12} + \Omega}, \quad g_2^* = \frac{\alpha}{W_{21} + \Omega}, \tag{5}$$

whose stability can be easily determined. Here $\Omega = \delta W_{21} W_{12}$ and $\alpha = \delta^2 W_{21} W_{12} - 1$. The eigenvalues associated to the Jacobi matrix for this system for $P_0^* = (0, 0)$ are

$$\lambda_{\pm} = -1 \pm \delta\sqrt{W_{12}W_{21}}, \tag{6}$$

and thus this point will be stable if $\delta\sqrt{W_{12}W_{21}} < 1$. There is an exchange of stability and P_1^* becomes stable when the previous condition does not hold (i. e. a transcritical bifurcation takes place). When self-interactions are also considered (i. e.$W_{ii} \neq 0$) several attractors can be present as a consequence of the competition between positive feedbacks and mutual inhibition. One particular case is given by networks such that the matrix of connections W is symmetric, of the form:

$$\mathbf{W} = \begin{pmatrix} \gamma & \beta \\ \beta & \gamma \end{pmatrix}, \tag{7}$$

with $\beta \in$ IR, and $\gamma > 0$. In other words, when there is self-activation by both genes plus cross interactions which can be positive or negative. The later is a very common situation in real morphogenetic processes and is strongly related with the process of competition between species in ecosystems. The stability analysis of this general problem can be performed by using the general Jacobi matrix:

$$\mathbf{L} = \begin{pmatrix} \alpha\delta/\epsilon_{12}^2 - 1 & \alpha\beta/\epsilon_{12}^2 \\ \alpha\beta/\epsilon_{21}^2 & \alpha\delta/\epsilon_{21}^2 - 1 \end{pmatrix} \quad (8)$$

where $\epsilon_{ij} \equiv 1+\alpha g_i^* + \beta g_j^*$. Fot $\beta > 0$ the mutual reinforcement between both genes leads to the same state (indicated as homogeneous in Figure 4). Here $g_1^* = g_2^* = [\delta(\alpha + \beta) - 1]/(\alpha + \beta)$ and it is stable (this point disappears at $\beta = (1 - \delta\gamma)/\delta$). For $\beta < -1$, the self-interaction is unable to sustain gene activity and it decays to zero. Finally, an interesting domain is observed for $(1 - \delta\gamma)/\delta > \beta > 0$, where three attractors are present (the previous one, where both coexist, and two exclusion points). In Figure 4(b) shows an example of the flow field for the 3-attractor domain.

We can see that there are three basins of attraction associated to each possible final state (fixed point). These results, in particular the presence of multiple attractors for some parameter ranges, are specially important within the context of development. In many cases the behaviour of cells that become differentiated is very similar to that of a switch. By depending on initial conditions or external perturbations, which might emerge from some other genes in the networks, the system can reach one or another basin of attraction and thus a different final state.

More importantly, it has been shown that some well-defined, small sets of interacting genes (so-called modules), are responsible for specific spatial patterns emerging in morphogenetic processes. As a consequence, not only single genes, but modules, can be the target of selection.

Random Networks

Beyond the specific wiring digrammes that can be considered

in small-sized genetic nets, the study of large-N nets has been dominated by randomly-wired systems. Here genes are connected at random, with an average number of z connections per gene.

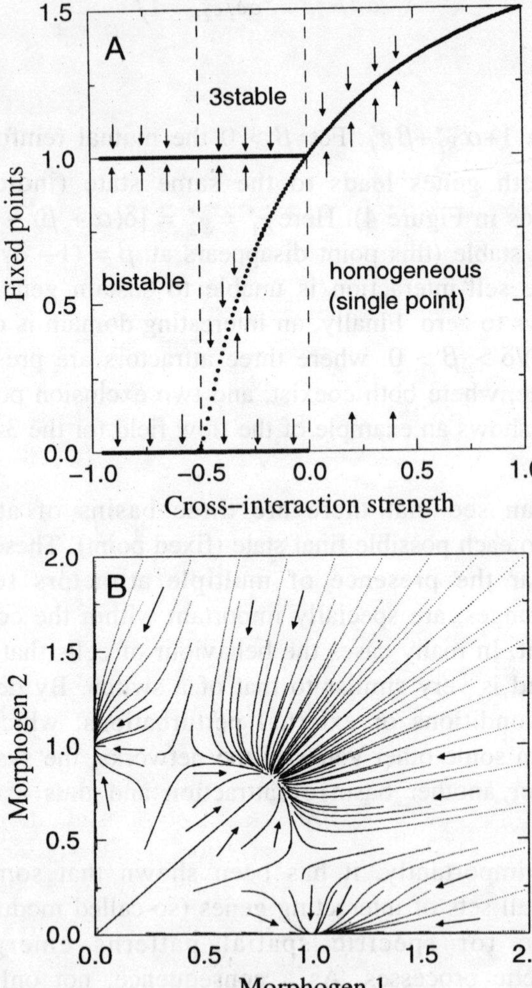

Figure 4. Multistability in gene network models

In this figure, (a) bifurcation diagramme for the two-gene network model with a symmetric matrix. $\delta = 2$ and $\gamma = 1$. Three basic domains are involved; (b) flow diagramme of the model for $\beta = -0.15$, in the three-attractor domain, indicated as 3stable in (a).

An extensive literature on random Boolean networks has shown that a number of generic features are characteristic of these nets as a consequence of the presence of phase transition phenomena in random graphs. In order to illustrate this idea, let us consider a graph $\Omega_{n,p}$ that consists of n nodes joined by links with some probability p. Specifically, each possible link between two given nodes occurs with a probability p. The average number of links (also called the average degree) of a given node will be $z = np$, and it can be easily shown that the probability $P(k)$ that a node has a degree k (it is connected to ã other nodes) follows a Poisson distribution,

$$P(k) = e^{-z} \frac{z^k}{k!}. \tag{9}$$

This so-called Erdos-Renyi (ER) random graph will be fairly well characterised by an average degree

$$\langle k \rangle = \sum_k kP(k) = z, \tag{10}$$

where $P(k)$ shows a peak. The distribution $P(k)$ is in this sense a single-scaled distribution and an example is shown in Figure 5(a). The ER model displays a phase transition at a given critical average degree $z_c = 1$. At this critical point, a giant component forms: for $z > z_c$ a large fraction of the nodes are connected in a single cluster, whereas for $z < z_c$ the system is fragmented into small subwebs. This type of random model has been used in different contexts, including ecological, genetic, metabolic, and neural networks. The importance of this phase transition is obvious in terms of the collective properties that arise at the critical point: communication among the whole system becomes possible, and thus information can flow from the units to the whole system and back.

Besides, the transition occurs suddenly and implies an innovation. No less important, it takes place at a low cost in terms of the number of required links (~ N). The ER model can be extended to directed graphs and has been analysed by Kauffman within the context of genetic regulatory networks. The N x N matrix W = $\{W_{ij}\}$ will have zN^2 nonzero elements, distributed at random. The probability that a gene is regulated by exactly k other genes will be then given by the distribution(9).

Beyond the specific timedependent features associated to the particular model chosen, one important characteristic of these systems is the presence of the percolation threshold: once a critical average connectivity z_c = 1 (the ratio of directed links to genes) is reached, the system becomes suddenly connected.Below the critical threshold the system is essentially disconnected and thus changes in a given gene cannot propagate to the rest of the system. The presence of the percolation threshold allows the system to exhibit a complex dynamical behaviour, including deterministic chaos, Figure 5(b).

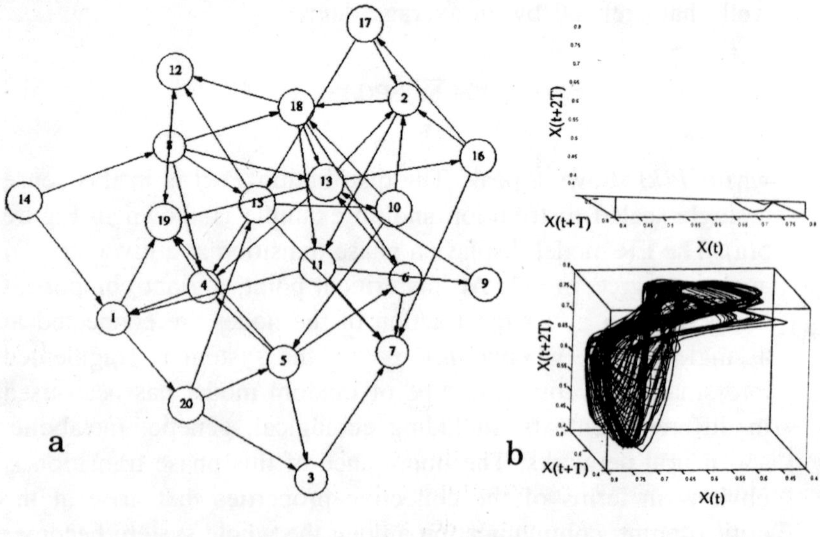

Figure 5. (a) An example of a directed random network with Poissonian structure.

One consequence of these models (but strongly tied to the topological properties of sparse random graphs) is that a high diversity of attractors compatible with a high degree of homeostasis seems to naturally emerge close to the percolation threshold. However, early evidence indicated that the degree distributions that characterise real genetic nets are far from Poissonian. Actually the topology of real networks strongly departs from the Erd¨os-Renyi scenario.

In figure 5, each node is a gene in a model gene network and arrows indicate the regulatory connections. This type of graph is characterised by an average degree *z*; together with the appropriate nonlinear coupling among genes, it can generate different types of dynamical patterns, including deterministic chaos. An example of the strange attractors obtained from these nets is shown in (b) in two different views.

Three Interconnected Levels

Gene regulation takes place at different levels and involves the participation of proteins. The whole cellular network includes three levels of integration:

— The genome, and the regulation pathways defined by interactions among genes;

— The proteome, defined by the set of proteins and their interactions; and

— The metabolic network, also under the control of proteins that operate as enzymes.

Unlike the relatively unchanging genome, the dynamic proteome changes through time in response to intra and extracellular environmental signals. The proteome is particularly important. Proteins unify genome structure on the one hand and functional biology on the other: they are both the products of genes and regulate reactions or pathways. Complicating the study of gene function is the fact that multiple proteins can arise from a single gene. In eukaryotic organisms, genes appear fragmented into pieces (exons) separated by non-coding domains (introns). After transcription, the resulting messenger RNA (mRNA) is generated by the excision and elimination of introns followed by the joining of exons. This process is called splicing.

Once the mRNA is formed, it will be translated into a protein by the translation machinery. A very important feature is that splicing can occur in different ways so that different sets of exons are joined together. In this way, different mRNA's (and thus different proteins) are produced. The combinatorial potential of this so-called alternative splicing is obvious. In some cases, thousands of different proteins are potentially available for a given gene. Alternative splicing expands genome complexity in an extraordinary fashion. Although the genomes of complex organisms might not strongly differ in terms of their number of genes, the underlying proteome complexity can be very different.

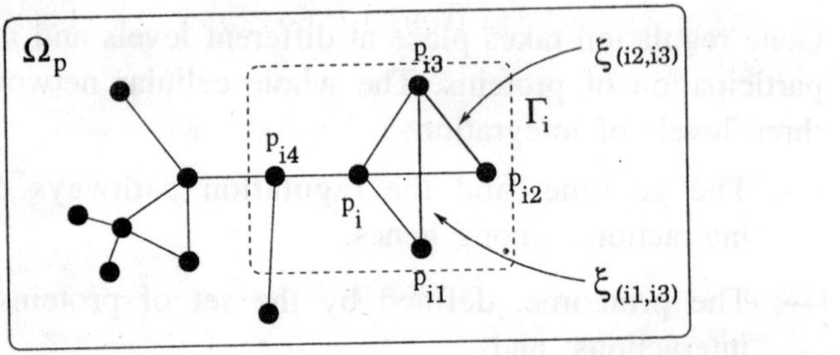

Figure 6. Measuring the clustering from a proteome graph Ω_p. Here each node (black circles) is a protein and physical interactions are indicated by means of edges connecting nodes.

Small World Graphs and Scale-free Nets

The analysis of the topological structure of protein interaction maps (in the budding yeast Saccharomyces cerevisiae and other simple organisms) revealed a surprising result: the proteinprotein interaction net shares some universal features with the topological organisation of other complex nets, both natural and artificial, ranging form technological networks, neural networks, metabolic pathways, and food webs to the human language graph. These studies actually offer the first global view of the proteome map and show that it strongly departs from the simple Erdos-Renyi scenario.

The first feature characteristic of the proteome map is that the probability $P(k)$ that a given protein interacts with other ã proteins has a scale-free (SF) nature, i.e. it follows a power law, $P(k) \sim k^{-\gamma}$, with a sharp exponential cut-off for large k. Thus most proteins have a small number of links with other proteins and a few of them are highly connected (hubs). Those last ones are likely to be very important to cell function. The second feature is the presence of the so-called small world (SW) property. Small world graphs have a number of surprising features that make them specially relevant to understand how interactions among individuals, metabolites, or species lead to the robustness and homeostasis observed in nature. The SW pattern can be detected from the analysis of two basic statistical properties of the network:

— the clustering coefficient C and

— the average path length $\bar{\ell}$.

The proteome graph (Figure 6) is defined by a pair $\Omega_p = W_p$, E_p), where $W_p = \{p_i\}$, $(i = 1, ..., N)$ is the set of N proteins (nodes) and $E_p = \{\{p_i, p_j\}\}$ is the set of edges/connections between proteins. The adjacency matrix ξ_{ij} indicates that an interaction exists between proteins $p_i, p_j \in W_p$ ($\xi_{ij} = 1$) or that the interaction is absent ($\xi_{ij} = 0$). Two connected proteins are thus called adjacent and the degree k_i of a given protein is the number of edges that connect it with other proteins. Let us consider the adjacency matrix and indicate by $\Gamma_i = \{p_j \mid \xi_{ij} = 1\}$ the set of nearest neighbours of a protein $p_i \in W_p$. The clustering coefficient for this protein is defined as the ratio between the actual number of connections between the proteins $p_j \in \Gamma_i$, and the total possible number of connections, $k_i (k_i - 1)/2$.(Figure. 6). Denoting:

$$\mathcal{L}_i = \sum_{j=1}^{N} \xi_{ij} \left[\sum_{k \in \Gamma_i} \xi_{jk} \right], \tag{11}$$

define the clustering coefficient of the i-th protein as

$$C(i) = \frac{2\mathcal{L}_i}{k_i(k_i - 1)}, \tag{12}$$

where k_i is the degree of the i-th protein. The clustering coefficient is defined as the average of $C(i)$ over all the proteins,

$$C = \frac{1}{N} \sum_{i=1}^{N} C(i). \tag{13}$$

The average path length $\bar{\ell}$ is defined as follows: Given two proteins p_i, $p_j \in W_p$, let ℓ_{ij} be the length of the shortest path connecting these two proteins, following the links present in the network. The average path length $\bar{\ell}$ is defined as:

$$\bar{\ell} = \frac{2}{N(N-1)} \sum_{i<j}^{N} \ell_{ij}. \tag{14}$$

For the ER graph, have a clustering coefficient inversely proportional to the network size, $C_{ER} \approx z/N$; this is a very small quantity, that tends to zero for large networks. The average path length, on the other hand, is proportional to the logarithm of the network size $\bar{\ell}_{ER} \approx \log(N)/\log(z)$. At the other extreme, regular lattices with only nearest-neighbour connections among units exhibit a long average path length. Graphs with SW structure are characterised by a high clustering $C \gg C_{ER}$, while possessing an average path comparable to an ER graph with the same average connectivity and number of nodes, $\bar{\ell} \approx \bar{\ell}_{ER}$.

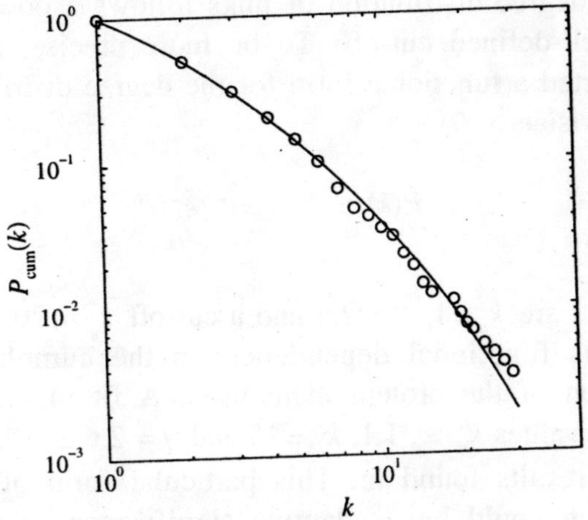

Figure 7. (b) Cumulated degree distribution for the yeast proteome map.

The degree distribution has been fitted to the scaling behaviour $P(k) \approx (k_0 + k)^{-\gamma} e^{-k/kc}$, with an exponent $\gamma \simeq 2.6$ and a sharp cut-off $k_c \simeq 15$.

The experimental observations on the proteome map can be summarised as follows:

— The proteome map is a sparse graph, with a small average number of links per protein. An average connectivity z ~ 1.9 – 2.3 was reported for the proteome map of S. cerevisiae. This observation is also consistent with the study of the global organisation of the E. coli gene network from available information on transcriptional regulation.

— It exhibits a SW pattern, different from the properties displayed by purely random (ER) graphs. In particular, reported the values $C = 2.2 \times 10^{-2}$ and $\bar{\ell} = 7.14$,to be compared with the values corresponding to an ER network with comparable size and average connectivity, $C_{ER} = 1 \times 10^{-3}$ and $\bar{\ell}_{ER} = 80$.

— The degree distribution of links follows a power-law with a well-defined cut-off. To be more precise, Jeong et al. reported a functional form for the degree distribution of S. cerevisiae

$$P(k) \simeq (k_0 + k)^{-\gamma} e^{-k/k_c}. \tag{15}$$

Parameters are $k_0 \simeq 1$, $\gamma \approx 2.4$ and a cut-off $k_c \approx 20$. In Figure 7 check this functional dependence on the cumulated degree distribution of the protein map[2] used. A fit to the form (15) yields the values $k_0 \simeq$, 1.1, $k_c \simeq 15$ and $\gamma = 2.6 \pm 0.2$, compatible with the results found in. This particular form of the degree distribution could have adaptive significance as a source of robustness against mutations. The highly heterogeneous character of these maps has important consequences within the context of molecular cell biology. It indicates that the evolution of proteome/genome complexity has been driven towards a well-defined topological pattern that provides the substrate for an extraordinary homeostatic stability against random mutational events.

Gene Duplication Models of Scale-free Proteomes

Several models have been proposed in order to explain the regularities displayed by the proteome map. These models of proteome evolution are based on a gene duplication plus rewiring process that includes the basic ingredients of proteome growth and intends to reproduce the previous set of observations.

The first component of the models allows the system to grow by means of the copy process of previous units. The second introduces novelty by means of changes in the wiring pattern, usually constrained to the newly created genes. This constraint is required if we assume that conservation of gene (protein) interactions is due to functional restrictions and that further changes in the regulation map are limited. Such constraint would be strongly relaxed when involving a newly created (and redundant) unit. The models proposed so far are intended to capture the topological properties of the proteome map.

No explicit functionality is included in the description of the proteins and this is certainly a drawback. But by ignoring the specific features of the proteinprotein interactions and the underlying regulation dynamics, one can explore the question of how much the network topology is due to the duplication and diversification processes. Rewiring has been used in dynamical models of the evolution of robustness in complex organisms. The proteome graph at any given step t (i.e. after t duplications) will be indicated as $\Omega_p(t)$.

The rules of the model, summarised in Figure 8, are implemented as follows. Each time step:

a) one node in the graph is randomly chosen and duplicated;

b) the links emerging from the new generated node are removed with probability δ;

c) finally, new links (not previously present) can be created between the new node and all the rest of the nodes with probability α.

Step (a) implements gene duplication, in which both the original and the replicated proteins retain the same structural properties and, consequently, the same set of interactions. The rewiring steps (b) and (c) implement the possible mutations of the replicated gene, which translate into the deletion and addition of interactions, with different probabilities.

Mean-field rate equation

Mean-field rate equation for the average connectivity: Since the model just presented has two free parameters, namely the deletion probability δ and the addition probability α, one preliminary task is to constrain their possible values by using the available empirical data. One average property that can be determined is the evolution of the average number of interactions per protein/gene through time, which can be compared with the evidence from real proteomes, as well as recent analysis of large-scale perturbation experiments.

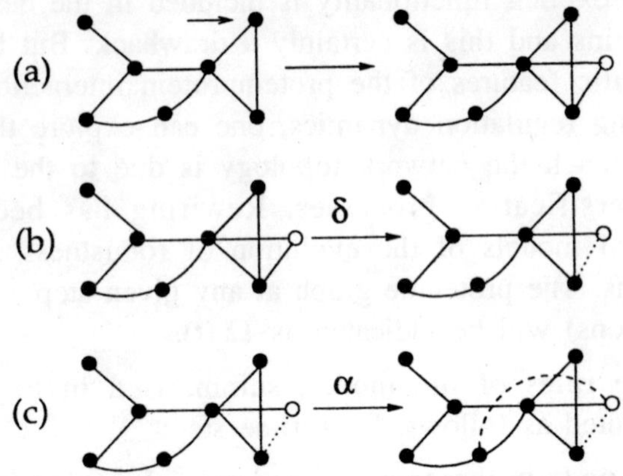

Figure 8. Growing network by duplication of nodes.

Int his figure, First (a) duplication occurs after randomly selecting a node (arrow). The links from the newly created node (white) now can experience deletion (b) and new links can be created (c); these events occur with probabilities δ and α, respectively.

Let us indicate by z_N and L_N the average connectivity of the system and its number links, respectively, when it is composed by N proteins. These magnitudes satisfy the relation $L_N = z_N{}^{N/2}$. It is easy to check that, at a mean-field level, that number of links L_N fulfill the following rate equation

$$L_{N+1} = L_N + z_N + \alpha(N - z_N) - \delta z_N, \tag{16}$$

where the last two terms correspond to the addition of links to a fraction α to the $N - z_N$ units not connected to the duplicated node, plus the deletion of any of the new z_N links, with probability δ. Using the continuous approximation

$$\frac{dz_N}{dN} \simeq z_{N+1} - z_N, \tag{17}$$

Eq. (16) can be written

$$\frac{dz_N}{dN} = \frac{1}{N}\left[z_N + 2\alpha(N - z_N) - 2\delta z_N\right], \tag{18}$$

whose solution is

$$z_N = \frac{\alpha}{\alpha+\delta}N + \left(z_1 - \frac{\alpha}{\alpha+\delta}\right)N^{\Gamma}, \tag{19}$$

where $\Gamma = 1 - 2(\alpha + \delta)$ and z_1 is the initial connectivity at N = 1. For any constant value of α and δ this model leads to an increasing connectivity through time. In order to have a finite z in the limit of large N, one possible solution is to impose an addition rate α that is a function of the size of the network, with the form

$$\alpha(N) = \frac{\beta}{N}, \tag{20}$$

where β is a constant. That is, the rate of addition of new links (the establishment of new viable interactions between proteins) is inversely proportional to the network size, and thus much smaller than the deletion rate δ, in agreement with the rates observed in. In this case, for large N, the differential rate equation (18) equation takes the form

$$\frac{dz_N}{dN} = \frac{1}{N}(1 - 2\delta)z_N + \frac{2\beta}{N}. \tag{21}$$

The solution of this equation is

$$z_N = \frac{2\beta}{2\delta - 1} + \left(z_1 - \frac{2\beta}{2\delta - 1}\right) N^{1-2\delta}. \tag{22}$$

For $\delta > 1/2$ a finite connectivity is reached in the limit of a large network,

$$z \equiv \lim_{N\to\infty} z_N = \frac{2\beta}{2\delta - 1}. \tag{23}$$

In order to reduce the number of independent parameters of the model, estimate the average degree z and the ratio of addition and deletion rates in the yeast proteome, α/δ to find a relation between β and δ, which, together with Eq. (23), yields a numerical estimate of β and δ. Since it is clear that this estimate is strongly dependent on the assumed value α/δ, considering a δ-dependent model and fixing the actual value of δ by comparing numerical simulations with experimental data.

Rate equation for the node distribution n_k

The rate equation approach to evolving networks can be fruitfully applied to the proteome model under consideration. This approach focuses on the time evolution of the number $n_k(t)$of nodes in the network with exactly k links at time t. Defining network by means of the set of numbers $n_k(t)$, the total number of nodes N is given by

$$N = \sum_k n_k, \tag{24}$$

while the total number of links is given by

$$L = \frac{1}{2}\sum_k kn_k. \tag{25}$$

Time is divided into periods. In each period, $t \rightarrow t + 1$, one node is duplicated at random, so that $N \rightarrow N + 1$ If, after each duplication, there is a probability δ to delete each link from the just-duplicated node, the probability of increasing the number of nodes at degree k, by direct duplication without link deletion, is given by

$$\text{Pr}_{\text{self,dup}}[n_k \rightarrow n_k + 1] = \frac{n_k}{N}(1 - k\delta). \tag{26}$$

On the other hand, a node of degree k can be created from the duplication of a node of degree $k + 1$ in which a link is deleted, contributing with a probability

$$\text{Pr}_{\text{above,dup}}[n_k \rightarrow n_k + 1] = \frac{n_{k+1}}{N}(k + 1)\delta. \tag{27}$$

The probability of degree change, from duplication of a node connected to a degree-k node, is given by:

$$\text{Pr}_{\text{other,dup}}[(n_{k-1}, n_k) \rightarrow (n_{k-1} - 1, n_k + 1)] = \frac{n_{k-1}}{N}(k - 1)(1 - \delta) \tag{28}$$

Finally, in the same period, proceed to add $N - k_d$ links with probability $\alpha = \beta / N$, where k_d is the connectivity of the just duplicated node. In the limit $N >> k_d$ we can simply consider the addition of $N\alpha = \beta$ new links to the graph. When this last step is performed with the correlated prescription given for the model (i.e. adding links from the duplicated node to the rest of the nodes in the graph), it leads to a nonlocal rate equation for the functions n_k. For the sake of simplicity, now consider the simpler case of a uncorrelated addition of links (new links created between any two nodes in the graph). However, it can be proved that both prescriptions lead qualitatively to similar results. The case of uncorrelated addition of links can be represented as the distribution of $2\alpha N$ new link ends among the N nodes in the network. This event contributes with a probability

$$Pr_{\text{add}}[(n_k, n_{k+1}) \to (n_k - 1, n_{k+1} + 1)] = \frac{n_k}{N} 2\alpha N = \frac{n_k}{N} 2\beta, \tag{29}$$

The probabilities (26), (27), (28), and (29) define the rate equation for the connectivity distribution

$$\begin{aligned} \frac{dn_k(t)}{dt} &= \frac{n_k}{N} + \frac{\delta}{N}[(k+1)n_{k+1} - kn_k] + \\ &\frac{1-\delta}{N}[(k-1)n_{k-1} - kn_k] + \frac{2\beta}{N}[n_{k-1} - n_k]. \end{aligned} \tag{30}$$

Since each time step a new node is added, Eq. (30) satisfies the condition

$$\frac{dN}{dt} = \sum_k \frac{dn_k(t)}{dt} = 1, \tag{31}$$

that yields the expected result $N(t) = N_0 + t$, where N_0 is the initial number of nodes in the network. In order to solve Eq. (30), impose the homogeneous condition on the population number

$$n_k(t) = N(t)p_k \simeq tp_k, \tag{32}$$

where p_k is the probability of finding a node of connectivity k, which assume to be independent of time. With this approximation, the rate equation reads

$$(k+1)\delta p_{k+1} - (k+2\beta)p_k + [(k-1)(1-\delta) + 2\beta]p_{k-1} = 0. \tag{33}$$

Eq. (33) can be solved using the generating functional method. Let us define the the generating functional

$$\phi(x) = \sum_k x^k p_k. \tag{34}$$

Introducing this definition into Eq. (33), an equation obtain for ϕ (x), whose solution is

$$\phi(x) = \left(\frac{\delta - x(1-\delta)}{2\delta - 1}\right)^{-2\beta/(1-\delta)} \tag{35}$$

Knowing ϕ (x) we can compute immediately the average connectivity

$$z = \sum_k k p_k \equiv x \frac{d\phi(x)}{dx}\bigg|_{x=1} = \frac{2\beta}{2\delta - 1}, \tag{36}$$

in agreement with the mean-field prediction of Eq. (23). On the other hand, performing a Taylor expansion of ϕ (x) around x = 0 we can obtain p_k as

$$p_k = \frac{1}{k!} \frac{d\phi(x)}{dx}\bigg|_{x=0} \tag{37}$$

Applying this formula to the function (35), and using Stirling's approximation for large k, we can obtain the asymptotic behavior of p_k, given by:

$$p_k \sim (k_0 + k)^{-\gamma} e^{-k/k_c}, \tag{38}$$

with

$$\gamma = -k_0 = 1 - \frac{2\beta}{1-\delta}, \qquad k_c = \frac{1}{\ln\left(\frac{\delta}{\delta-1}\right)}. \tag{39}$$

As we can observe from the previous result, recover the same functional form experimentally observed in. However, it is important to notice that for all the parameter range in which the exponential cut-off kc is well-defined, we obtain a value of the degree exponent, as given by Eq. (39), that is $\gamma \leq 1$. The same result holds when considering the rate equation for the correlated model, in which the link addition is fully correlated with the new duplicated node. This result is unsatisfactory, because it does not correspond with the results from numerical simulations of the model. This discrepancy is explained by the fact that the $N \rightarrow \infty$ solution presented has only meaning for $\delta > 1/2$ (Eq.36).

Yet the master equation was defined on the basis of an independent-event approximation that only makes sense for $\delta << 1$. The master equation itself should become valid for $\delta \rightarrow 0$, but then the convergence results assumed at $N \rightarrow \infty$ seem questionable, as indicated by the fact that we get an analytic, but negative, z. There is, however, something qualitative still to be learned from these equations, in the neighborhood of $\delta \sim 1/2$ small β. This is a neighborhood where the convergence results at large N still give sensible answers, even if they are not quantitatively correct due to marginal approximations in the underlying master equation.

Yet at the same time, since this is the smallest value of δ where we can get answers, it is the one where the master equation constructed is likely to be the best approximation to the much more complicated true equation (one with frequent coupled events).

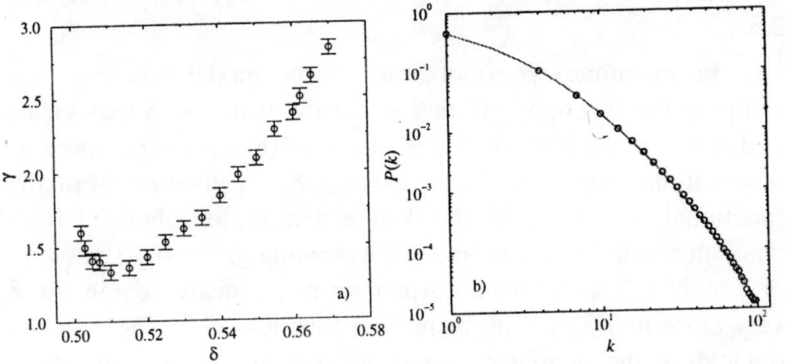

Figure 9

In figure 9: a) Degree exponent γ as a function of the deletion rate δ from computer simulations of the proteome model with average connectivity z = 2.5. Network size $N = 2 \times 10^3$. The degree distributions is averaged over 1000 different network realisations. b) Degree distribution for the same model, δ = 0.562, averaged over 10000 different network realisations. The distribution can be fitted to the form $P(k) \approx (k_0 + k)^{-g} e^{-k/kc}$, with an exponent $\gamma = 2.5 \pm 0.1$ and a sharp cut-off $k_c \simeq 37$.

Numerical simulations

The proteome model depends effectively on two independent parameters: the average connectivity of the network z and the deletion rate of newly created links δ, being the addition rate β computed from Eq. (23). The average connectivity can be estimated from the experimental results from real proteome maps. If value $z \simeq 2.40$, as a safe estimate, one can impose the value z = 2.5, and consider values $\delta > 1/2$, in accordance with Eq. (23). In spite of the drawbacks of the analytical study of the model, rate equation for the node distribution n_k, one should expect the model to yield, for each value of γ, the functional form Eq. (38) for the degree distribution, with a degree exponent

T which is a function of δ (for a fixed average connectivity z = 2.5).

From numerical simulations of the model one can then compute the function $\gamma(\delta)$ and select the value of δ that yields a degree exponent in agreement with the experimental observations. Fig 9(a) shows values of γ estimated from the functional form (38) for the degree distribution obtained from computer simulations of model, averaging over 1000 network of size $N = 2 \times 10^3$ nodes. Apart form a concave region for δ very close to 1/2, γ is an increasing function of δ. The value of δ yielding the degree exponent closest to the experimentally observed one is then

$$\delta = 0.562. \tag{40}$$

In Figure 10. (a) show the topology of the giant component of a typical realisation of the network model of size $N = 2 \times 10^3$.

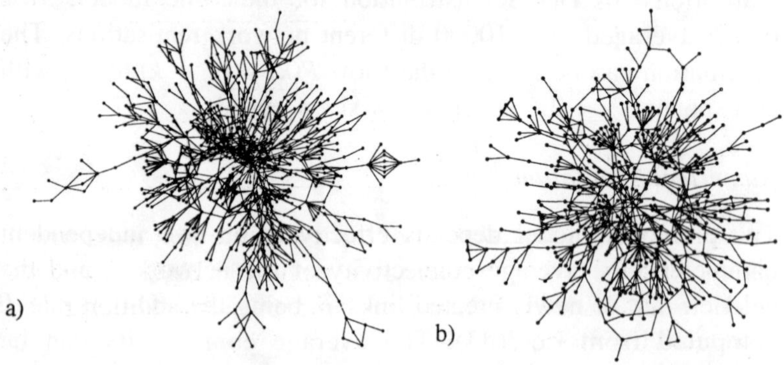

Figure 10. a) Topology of the giant component of the map obtained with the proteome model with parameters with k = 2.5 and δ = 0.565. Network size N = 2 x 10³. b) Topology of a real yeast proteome map obtained from the MIPS database.

This Figure clearly resembles the giant component of a real yeast networks, as we can see comparing with figure 10(b); we can appreciate the presence of a few highly connected hubs plus many nodes with a relatively small number of connections, in close resemblance of the real proteome map.

On the other hand, Figure 9(b) shows the connectivity $P(k)$ obtained for networks of size , averaged of 10000 realisations, for $\delta = 0.562$. In this Figure the resulting connectivity distribution can be fitted to a power-law with an exponential cut-off, of the form given by Eq. (38), with parameters $\gamma = 2.5 \pm 0.1$ and $k_c \simeq 37$.

Table 1 reports the values of z, γ, C and ℓ obtained for the proteome model, compared with the values reported for the yeast S. cerevisiae, and those corresponding to an ER random graph with size and average connectivity comparable with both the model and the real data.

Table 1. Comparison between the observed regularities in the yeast proteome

	Wagner's data	Jeong's data	Proteome model	ER model
z	1.83	2.40	2.4 ± 0.6	2.50 ± 0.05
γ	2.5	2.4	2.5 ± 0.1	–
C	2.2 x 10^{-2}	7.1 x 10^{-2}	1.0 x 10^{-2}	1x 10^{-3}
ℓ	7.14	6.81	5.5 ± 0.7	8.0 ± 0.2

Wagner, those calculated from the proteome map used by Jeong et al. the model predictions with $N = 2000$, $\delta = 0.562$ and $z = 2.50$, and a ER network with the same size and connectivity as the model.

All the magnitudes displayed by the model compare quite well with the values measured for the yeast, and represent a further confirmation of the SW conjecture for the protein networks.

5

Modelling of Biological Networks

Modelling, reverse-engineering and analysis of macromolecular networks has spurred increasing interest in the computational biology and the biostatistics communities. With the availability of complete genome sequences and high-throughput, post-genomics experimental data, the last 5 years have witnessed a growing interest in the study of networks of macromolecular interactions. During the last few years, modelling efforts have targeted several distinct types of networks at the molecular level: gene regulatory networks, metabolic networks, signal transduction networks or protein-protein interaction networks, not to mention networks of interactions that are not restricted to a cell (intercellular communications) or take place at an altogether different level of detail (immunological networks, ecological networks).

A major challenge consists in identifying with reasonable accuracy those complex macromolecular interactions that take place at different levels from genes to metabolites through proteins. Once identified, a network model can be used to simulate the process it represents, or for a variety of analyses, ranging from statistical properties of its topology to predictions of features of its dynamic behaviour, or even prediction of cellular phenotypes.

Network Topology Analyses

The first category of properties of interest in biological networks are those related to their (static) network structure. Such topological analyses are most meaningful when applied to large ('genome-scale') networks, the aim being to identify statistical properties that can be interpreted as'traces' of underlying biological mechanisms or design principles, related for instance to their dynamics, to their evolution (i.e. likely scenarios for the evolution of a network exhibiting the observed property or properties). Those analyses that are motivated by the search for insights into network dynamics focus on network structure mostly because large-scale data on network dynamics is not yet available. They can provide valuable insight insofar as the interpretative leap between static structure and dynamic behaviour is performed carefully.

A gene regulatory network is viewed as a directed graph : a pair (V,E) where V is a set of vertices and E a set of directed edges, i.e. pairs (i, j) of vertices, where i is the source vertex and j the target vertex. Vertices of the graph represent genes, edges represent regulatory influences. In some cases, it may be preferable to work with undirected graphs instead,

— using a more or less implified set of entities and reactions, e.g. choosing whether to take into account mRNA and protein degradation,

— including delays to account for transcription, translation or diffusion time

— using more or less detailed kinetics, i.e. specific forms of f_i

Systems of differential equations as a modelling framework for biological networks presents two major drawbacks. Each equation in the model requires the knowledge of one or several parameter values (thermodynamic constants, rate constants), which is out of the present reach of high-throughput data production techniques. It is thus difficult to instantiate models of large net-works directly, and reverse-engineering techniques are limited in how much information they allow to extract from limited datasets. Moreover, deriving meaningful dynamical properties of large differential equations system is a challenge : the f_i being nonlinear, analytical solutions are not known in the general case.

So far these systems have been mainly used for numerical simulations within given parameter ranges (realistic or not), possibly complemented by bifurcation analysis, rather than submitted to analytical approaches. These limitations have motivated two main tracks of investigation on alternative modelling frameworks for biological networks : simplified kinetic models on one hand, and discrete models on the other hand. Simplified continuous frameworks include piecewise-linear differential equations, a special case of rate equations where the response of a gene to regulatory stimuli (the function fi) is approximated by a step function.

Linearity facilitates the analytical treatment of some dynamical properties, such as steady states. Systems of piecewise-linear differential equations can also be analysed qualitatively by discretising and recasting them within the framework of qualitative differential equations, where variables and their derivatives take qualitative (discrete) values and functions f_i are abstracted into sets of qualitative constraints. Several discrete modelling frameworks have been proposed, each with a specific tradeoff between the level of detail of its chosen observables and the type of analyses that it enables : boolean networks, generalised logical networks (a generalisation of boolean networks that increases biological realism by allowing variables to have more than two values and using asynchronous transitions), petrinets, process-algebra, rule-based formalisms.

Constraint-based modelling is a framework dedicated to the modelling of metabolic processes at steady state : a global state of the metabolic network is defined as a distribution of fluxes within the network reactions. It emerged in the 90s as a simplification of kinetic models, and was developed to allow tractable modelling of genome-scale metabolic networks. The steady-state hypothesis positions the framework at a level of detail intermediate between description of static network structure and representation of network dynamics. It is designed to represent incomplete information, yet to allow some prediction of metabolic behaviour.

The focus, rather than being on fully instantiated descriptions of the system's behaviour, is on sets of such descriptions, i.e. sets of flux distributions compatible with a set of constraints representing the current knowledge on the structure of the network, on thermodynamic and kinetic parameters, and on input/output relationships of the network with its environment. The solution set can be refined incrementally as new constraints are added, ensuring some robustness in structural analyses and metabolic behaviour predictions with respect to modifications of the model.

Identification of Model

A machine learning problem: Once a formal framework is defined to describe models of biological networks, the question of how to choose parameters arises. Various works have shown that this identification problem can be expressed in the framework of machine learning. Given a family of mathematical models of gene interactions and a set of observations, learning consists here in optimising the parameters of the model in such way that it captures the observed behaviour of the true system. The ability of the instantiated model to be used in prediction is referred as the generalisation property.

A model is able to generalise if learning ensures a trade-off between a good fit to the data and simplicity of the model. Solving a learning problem leads to three key questions : the representation problem, the optimisation problem and the validation problem. The representation problem concerns mostly the choice of the formalism in which data and the model are going to be expressed, and the method to encode them into this formalism. Both symbolic and numerical learning leads to an optimisation problem whose nature is combinatorial (for symbolic learning) and numeric (for statistical learning). Statistical approaches generally use maximum likelihood criteria penalised by a parsimony constraint.

Combinatorial approaches are solved using heuristics to ensure a large exploration of the models spaces. At last validation is required to identify how one can trust the inferred model. In this area statistical approaches becomes also more complex, the door to new formalisms that couple high level descriptions with a probabilistic framework. Another variations on bayesian networks models is the so-called 'module networks' approach. Module networks introduced in have been proposed as bayesian networks with a special structure, where the variables sharing the same parents are gathered into a so-called module, i.e. a set of genes that appear to be co-regulated in some experimental conditions.

Elucidating which are the genes that belong to the same modules and what are the conditions under which these regulations occur can be solved using a Expectation Maximisation (EM) based algorithm that starts from relevant initialisations. Validation can consist in the comparison between the inferred structure and the true structure. The simplest way consists in a comparison between the inferred network with other sources of knowledge. Precision and recall measures, ROC curves have also been proposed to evaluate the power of learning algorithms. However all these static approaches are not able to discover circuits in a graph interaction. The reason is that without considering time, it is not possible to elucidate feedback interactions that can only be observed an through time.

Dynamic approaches aim at identifying the dynamics of the system imple-mented by a biological network while extracting the structure. Only discrete-time models are considered for learning since experimental data come from discrete-point measurements. In the area of genetic networks, the available data take the form of gene expression kinetics measured after some perturbation of the studied organism. While these data are more expensive to generate than static data, a few subsets exist and mainly concerns model organisms such as bacteria or yeast. Modelling dynamics of a network can serve both exploratory and explanatory goals.

A long-term goal is of course to exploit these models in simulation and prediction for drug-design and therapeutical targeting. Dynamic models that have been considered for learning include Boolean networks, artificial recurrent neural networks, dynamic bayesian networks including state-space models. Learning in boolean networks has first been tackled with combinatorial algorithms and then renewed by using a randomised algorithm.

However the best way to reduce complexity of the problem is to reduce the class of boolean functions as proposed in with the so-called chain functions. Promising new directions have also been introduced by with the introduction of Probabilistic Boolean networks and learning algorithms devoted to their reverse engineering.

Modelling and Network Organisation

Modelling and analysis are uniquely suited to a wide variety of applications. Models can be used to test specific hypotheses, for instance, about how a system is structured or how it functions. They can be used to make predictions, which can then be tested with appropriate experiments. Models can be used in a more exploratory manner, for instance, to discover the types of properties that might emerge from integration of parts with specific properties in specified ways. While experiments can also be used to test hypotheses, make predictions and explore, mathematical models explicitly represent components and their interactions in a controllable, manipulable environment (the computer) which allows calculation of how these things change through time and space (if those are included) and observation of every element and relationship in the model.

The analogous experimental measurements would frequently be difficult or impossible and sometimes unethical.

In addition to these specific applications in research, models are also excellent aids to communication and teaching. For several decades, models and theory were mostly the domain of applied mathematicians, physical scientists and engineers, many of whom worked rather independently of experimental science and the work remained somewhat obscure and theoretical. With the broad availability of computers and IT infrastructure that has emerged in the last several decades, the use of modelling and theory in biological research has expanded greatly, as has the size of the models being developed. Historically, much modelling was used to study and interpret what could be directly observed in the laboratory, namely, functions of cells, tissues, organs and organism physiology.

With the advent of molecular biology and ensuing capabilities in genomics, proteomics, and so forth, mathematical models are now being used extensively to study intracellular molecular networks such as kinase cascades and metabolic pathways, as well as gene regulatory networks. These intracellular molecular networks are a primary subject of network organisation analysis, along with epidemiological networks, and structural networks such as lung airway, vascular, and neural network topologies.

Mathematical Model

A model is a set of structured assertions that specify the interactions among entities of a network. An example of a model that illustrates this definition is given in Figure 1. The entities in a model can be many different things, such as properties of specific biological elements (e.g., molecular concentration, cellular density, or organ volume) or specific physiologic characteristics (e.g., blood pressure, heart rate, or weight). Likewise the interactions that are specified among the entities can represent various processes such as molecular reactions, binding of a molecule to a cell-surface receptor, subsequent stimulation of that cell, etc.

From a model one can calculate how the entities change over time and/or through space or their value at steady state. Mathematical models of biological systems can take many different forms, and the appropriate types of mathematical equations are highly dependent on the problem one is attempting to represent with the model. Many models utilise ordinary or partial differential equations to represent continuous, deterministic systems and partial differential equations if space or mechanics are involved.

Reaction system

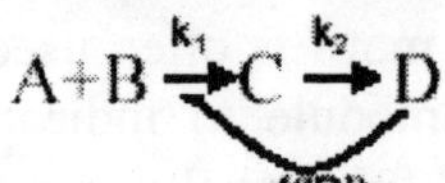

Model of dynamics

$$\frac{d[A]}{dt}=\frac{d[B]}{dt}=-k_1 f([D])[A][B]$$

$$\frac{d[C]}{dt}=k_1 f([D])[A][B]-k_2[C]$$

$$\frac{d[D]}{dt}=k_2[C]$$

Model solution

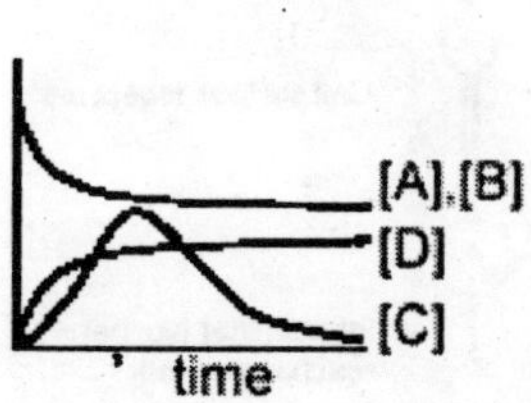

Figure 1. Schematic of a chemical reaction system where chemical species A and B react together to form C, which further changes into D, and D inhibits the first reaction. The equations of a mathematical model of that system using ordinary differential equations (ODEs) are shown along with a graph of a simulation of that model, assuming A and B start with equal, non-zero concentrations while C and D are initially zero. Brackets around a chemical name indicate concentration.

Various methods are used to include stochastic or probabilistic properties of the system including Langevin dynamics and the more physically rigorous chemical master equation. Discrete methods like particle/molecular dynamics, cellular automata and agent-based models are utilised where the actions of individual elements of a system, rather than the population behaviour, is of interest.

Network Organisation Analysis

Biological networks are not randomly structured, but rather that various parts of the network have structures that provide specific functional units, and those structures can be found in multiple parts of the network. Although universal definitions are not agreed upon, these subunits with the same structure are often called motifs or modules, where the term motif is often used to indicate the smallest repeated unit and module to indicate a collection of units that form a "separable" functional group-that is, a group of processes that together create some well-defined behaviour that is a pure input/output function and is not otherwise affected by inclusion of the network-although those definitions are not universal.

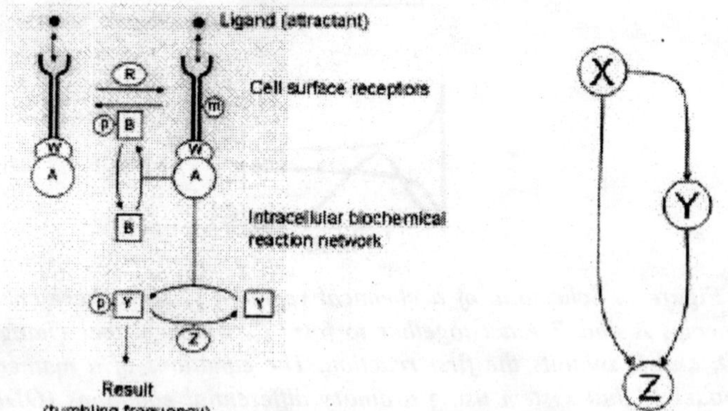

A. B.

Figure 2. Example of (A) a molecular interaction network involved in bacterial chemotaxis and (B) a single motif, the feed-forward loop, frequently found within intracellular molecular networks.

A module in the context of gene expression as a group of genes that tend to respond in a joint manner, i.e., through temporally coordinated gene expression. A module in the context of network topology as a discrete group of interconnected elements that is abstracted from the topology of the network. A module functionally as a set of genes and their products which, as an emergent consequence of their interactions, perform some task nearly autonomously.

Figure 2 illustrates molecular networks and motifs. These smaller units are thought to derive from functional need of the organism as well as fundamental physical principles (e.g., thermodynamics). The analysis and theory of network organisation focuses on discovering underlying principles and motifs of systems and networks. Many different approaches to network organisation analysis are being used in biology and various reviews are available.

Involvement of Industry in Systems Biology Efforts

The interest and involvement of industry in systems biology efforts that include modelling and network organisation work are significant although hardly ubiquitous, and quite variable between regions. The industry of concentration is pharmaceutical and biotechnology, although applications in the nutraceutical, agricultural supply/chemical and bioprocess industries also exist. Here activities within the industry as well as the relationship between industry and academia are summarised.

R&D Using Modelling

Companies using mathematical models of biological systems are treating them essentially as in silico laboratories that complement the experimental laboratories. Models are particularly suited to tracking the states of and relationships among numerous elements through time and space, especially for large, complex systems as found in biology and medicine. Applications of systems biology modelling within the pharmaceutical industry vary from drug target identification to clinical trial design and analysis. Models of intracellular biochemical networks are typically utilised to investigate how modulation of that network, by agonising or antagonising network components, affect the associated cell function.

The direct relationship of that cell function to a particular disease is disease-specific, for instance proliferation of a cancer cell is directly related to disease outcome (tumor size, say), whereas modulation of an inflammatory cell within the asthmatic airways still requires extrapolation to relevant clinical endpoints. Models of whole organs in which a disease is isolated, like the heart for cardiac arrest or arrhythmia, or of multicellular/tissue/organ systems that include relevant clinical endpoints, bridge cellular function to relevant clinical outcomes much as an experimental animal model would. Given that drug targets are nearly always molecular, modelling within the pharmaceutical industry is most frequently focused at the level of intracellular biochemical networks.

Several companies in the U.S. that are focused on development and use of such intracellular network models include Gene Network Sciences (U.S.), Merrimack Pharmaceuticals (U.S.), Genomatica (U.S.) and Physiomics (U.K.). Gene Network Sciences uses inference modelling and mechanistic simulation of intracellular biochemical and gene networks related to cell cycle to do research on cancer and its treatment. More recent directions include cardiac electrophysiology. Physiomics also focuses on modelling cell cycle control. EGF receptor dynamics and downstream signaling associated with various cancers are the subject of modelling at Merrimack Pharmaceuticals.

In contrast, Genomatica utilises models of microbial and yeast cell metabolism to improve bioproduction of chemicals and proteins, among other applications. Multicellular/tissue/organ network function has also attracted the application to pharmaceutical development. Entelos, Inc. (U.S.) utilises dynamic ODE models of the biological systems involved in specific diseases to evaluate drug targets, select lead compounds, predict biomarkers, and design clinical trials. Current areas of work include model representations of human diseases (asthma, hematopoiesis, obesity, rheumatoid arthritis, and type 2 diabetes), and animal models of human disease (type 1 diabetes).

Larger pharmaceutical companies that utilise biological systems modelling in some of their R&D activities include Pfizer, AstraZeneca, Hoffman-La Roche, Johnson & Johnson Pharmaceutical Research Division, GlaxoSmithKline, Novartis, Organon, and Bayer. This list is probably not exhaustive. In some of these companies the inclusion of modelling is quite extensive, while in many it is often confined within a single therapeutic area or a single group that works with multiple experimentalists. There seem to be fewer companies primarily devoted to using modelling as a primary R&D method in other areas of the world.

In addition to Physiomics mentioned above, another modelling-focused company is Optimata in Israel, whch utilises modelling to optimise drug dosing and schedules, in particular for cancer treatment. While companies are notably less forthcoming than academic researchers with making their models and research results public, there is emerging evidence that modelling is significantly impacting pharmaceutical development.

Relationships Between Industry and Academia

Close relationships between industry and academia were particularly obvious in Japan in comparison to the U.S. and Europe. Many Japanese academic researchers stated that the government research funding agencies strongly encouraged collaborations with industry and the transfer of technologies to industrial concerns, either through the start up of new companies or to established companies. In numerous laboratories there was active involvement in the research by industry staff in residence. In contrast, in Europe the relationship of industry to academic research was at the level of a few collaborations (without exchanging personnel) and the encouragement by funding agencies to transfer findings and technology developments to industry.

The recent large collaborative projects in Germany and the EU described above were both influenced by the interests of the pharmaceutical industry, although in the Hepatocyte Project there is no actual industry involvement. The EU Biosimulation project is too new to know how involved the corporate partners will become beyond providing funding. In the United States, academic-industry collaborations are common, although it is uncommon for industry personnel to spend time in an academic lab. The reverse is probably more likely. It has become the norm for universities in the U.S. to patent research findings and then license them to companies or for the inventors to start new companies to commercialise the inventions.

Infrastructure Supporting Modelling and Network Organisation Analysis

Throughout the U.S., Europe and Japan, the researchers found significant efforts devoted to the development of software platforms in which to build and simulate mathematical models of biology (and sometimes more general) systems. Reasons given by researchers for developing new platforms included the need to have faster simulation capabilities, improved usability, and features specific to the biologic system being modeled by the developer's research group. Usability issues were mentioned with particular attention to making models and simulation more accessible to non-expert users, although technical usage was also mentioned. The latter includes such things as ease of specifying model equations (e.g., specifying equations by drawing structured diagrams rather than typing them), input and modification of parameter values, specification of protocols to simulate, and storage/retrieval of simulation specifications and results.

Features desired were typically specific mathematical methods such as those to handle stochastic processes or finite element algorithms, as well as analytical methods such as parameter optimisation or sensitivity analysis. Some modelling software is available commercially, the most commonly used (at least in the U.S.) general purpose numerical computing platform being MatLab by The MathWorks, followed perhaps by Mathematica (Wolfram Research). Matlab and Mathematica both have announced special tools for systems biology.

More specialised commercial software includes Berkeley Madonna, and a number of pharmacokinetic simulation packages. Benefits of such commercial platforms are their formal quality assurance/quality control (QA/QC) processes as well as formal means for reporting bugs and requesting new features. The disadvantage is that they are not free, although the cost of most commercial packages tends to be nominal for academic use. Many of the mentioned packages are not specialised for biological systems, are not well linked to biological databases and data sources, and are not user-friendly for the biological user.

For the many platforms being developed noncommercially within research groups, the most common method of dissemination (when available) is via the group's web site. The advantage of development within research groups is that the software features and user interface can be closely guided by end-users within the group. Some disadvantages, however, are that QA/QC procedures for such software are frequently unclear and likely absent in many cases and "services" to users outside the developer's group, such as ways to report bugs and request new features, are not always clear and response by developers not assured.

The fact that so many new packages are in development strongly suggests that those available, commercial or otherwise, are not fully meeting significant needs of the biological modelling community. While no one piece of software will meet the needs of all modelling efforts, the community would be wellserved by a national or international resource devoted to making broadly applicable platforms widely and freely available, as well as supporting maintenance and expansion. Such a resource should reduce the ongoing proliferation of somewhat duplicative and quite expensive software development.

The Bio-SPICE programme funded by Defense Advanced Research Projects Agency (DARPA) for the last three years is an example of such a programme although it focused primarily on initial development for an open-source infrastructure for integrating such software. The funding for this project ends in 2005 and currently there are no allocated resources from DARPA or elsewhere to continue funding to support dissemination of the resulting software or its continued development.

Sharing of Models

Historically models have been shared through literature publication. As modelling has become a more common way to do biologic research and the models themselves have become larger, this method has become less satisfactory to many and a strong desire to more easily share models in electronic form between research groups has grown. While some researchers simply want to use or modify published models without having to re-generate computer code to do so, others also want to integrate others' models with their own to create models of larger biological systems.

The main difficulty in sharing models electronically is that models are encoded for specific software and hardware platforms that aren't universally compatible with software or hardware in other labs. To alleviate this problem, two main international efforts are underway to better enable model sharing, namely, the development of markup languages that encode mathematical equations typically used in models of biological systems. For encoding models, Systems Biology Markup Language and CellML are two major languages under development.

The idea is to create a language analogous to HTML (hypertext markup language), the common encoding language for the web. As long as one has an HTML "decoder" on one's computer, e.g., a browser such as Microsoft Internet Explorer, then one can interpret the text, pictures, etc. encoded in an HTML file and the same file can be viewed on all computers.

SBML is focused on language to encode math that describes biochemical reaction networks while CellML is focused on describing cellular components and compartments as well as biochemical reaction networks.

SBML began in Japan and at CalTech and has since become an international collaboration effort with funding from the U.S., Japan and the U.K. Development efforts on CellML are based in New Zealand. At least one other biological modelling-focused markup language is being developed by Satoru Miyano's group in Japan, labeled Cell System Markup Language, although it does not appear to be publicly available yet.

Part of the effort associated with both SBML and CellML is the development of a public repository of models online in the markup language. Authors of models are encouraged to deposit a version of their model on the web site, available for download by others. While none of these markup languages are as yet accepted as the standard, researchers generally acknowledge SBML as being the most frequently used. Certain U.S. funding agencies, including the National Science Foundation (NSF), have taken the position that models developed with their funding must be made publicly available in SBML.

Need of Modelling Efforts

There are several needs and deficits of the modelling efforts in systems biology and makes several recommendations to alleviate these. First and foremost, the researchers notes the need for substantially greater and more widespread integration of modelling and experimental programmes. Currently the majority of modelling and experimental efforts on a given subject are performed somewhat or completely remotely from each other. This decreases the benefit of both to each other, and slows progress in developing new understanding in many fields.

Data from diverse laboratories and with diverse protocols are often found to be difficult to compare when placed in model context implying both that models can help ensure consistency among datasets thus preventing spurious conclusions about the significance of a particular observation and that data quality control is even more important than model quality control. While it may seem from the highlighted examples above that modelling work is commonly integrated with experimentation, the examples were selected in part because they demonstrated how well-integrated programmes provided unique insights into the subjects of study.

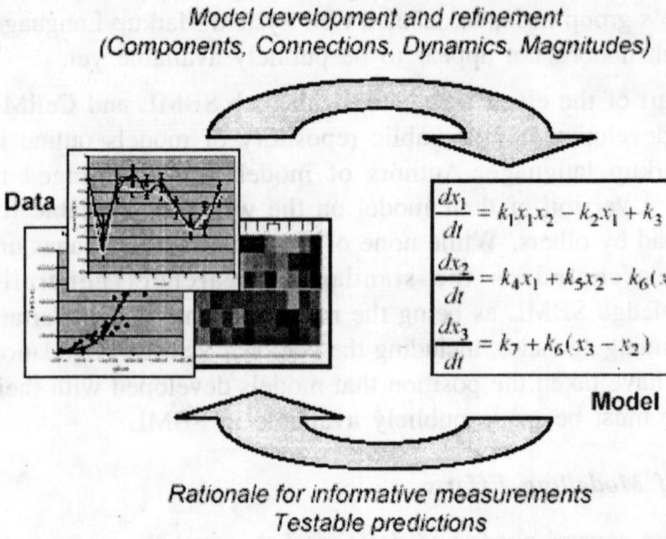

Figure 3. With the purpose of increasing understanding of a biological system function, one needs a set of data to develop the first model. That first model can then be simulated or analyzed to pinpoint uncertainties that are important to the system and then recommend a new set of experiments to measure relevant quantities to reduce those uncertainties. The new data can be used to revise the model. The model at various stages of iteration can also be used to test hypotheses about the system's function or means of modulating that function, and interesting predictions from those tests can then also be verified (or refuted) experimentally.

The reason that such tight integration is promoted by the researchers is because models can't be developed or tested without experimental data, and the experimentation that provides the necessary data is often not obvious without the model guiding its design. The experiment-model iteration paradigm that is most productive is illustrated in Figure 3. Closely related to the need for experimental-modelling integration within research endeavors is a need for improved means of comparing experimental data and modelling results.

Many software platforms for modelling currently don't support easy representation of experimental data, for instance, so both the data and modelling results have to be exported to a third software to allow the comparison. Another major need is for a means of disseminating and maintaining good cheap or free software appropriate to modelling of various problems nationwide and internationally.

6

PATHWAY ANALYSIS IN METABOLIC NETWORKS

Networks composed of the building blocks of the cell and their interactions are qualitative descriptions of the working of the cell. Such networks can be modeled as graphs. Metabolic networks are typical examples of such networks. They are composed of biochemical entities participating to reactions as substrates or products. Such a network can be modeled as a bipartite digraph which nodes are the biochemical entities and reactions and edges are the substrate or product link between an entity and a reaction.

Pathways are specific subsets of a metabolic network which were identified as functional processes of cells. As these pathways are known to be working processes of the cell, they can be used to study the metabolic network. One type of metabolic network analysis consists in finding simple paths in the metabolic graph.

A potential application is the explanation of DNA chip experiments using a CSP able to solve pathway discovery problems. The study of the metabolic network is constantly evolving and most of the problems are solved with dedicated algorithms. This dedicated approach has the benefit of yielding very efficient programmes to solve network analysis problems. This approach however has the drawback that it cannot be easily adapted to solve other problems or easily combined to solve combinations of various analyses.

This declarative paradigm allows to easily adapt programmes or combine different programmes. In order to provide a high level modelling language and as the data and results are graphs, they defined a graph computation domain for constraint programming.

Stoichiometric Network Analysis

Stoichiometric network analysis (SNA) of biochemical reaction systems has become an important approach for understanding the functionality of metabolic networks reconstructed by genomic and biochemical data. SNA relies on the mass balances of metabolites and the assumption of metabolic pseudo-steady state leading to the homogeneous system of linear equations set up by the stoichiometric matrix *N* (m rows: metabolites; *q* columns: reactions) and the vector *r* of net reaction rates (q elements):

$$0 = rN \tag{1}$$

Whereas metabolic flux analysis focuses on single flux distributions, Metabolic Pathway Analysis studies the complete set of admissible flux distributions. A promising approach is the concept of elementary flux modes (EFMs), which are unique for a given metabolic network and can be considered as nondecomposable steady state flux distributions using only a minimal set of reactions.

A pathway-oriented view of the metabolism by EFMs helps to understand the "logic" of network routing, and to study a number of aspects of metabolic networks.

Flux Modes and Combinatorial Complexity

So far, most applications of EFMs avoid to analyse more complex networks (e.g. covering all the main routes of the catabolism) due to the enormous combinatorial explosion of the number S of EFMs. It is therefore worth studying the combinatorial complexity of metabolic pathways.

Understanding the EFMs as subsets of the columns (reactions) in N being minimal dependent we can determine the maximal number $S_{\max}$ of EFMs as follows: Assuming that N has more reactions than metabolites ($m<q$) and full rank m, the maximal number of EFMs occurs when all reactions are reversible, and each possible subset of columns of N having m elements is independent, i.e. they all represent a basis of the columns of N.

In this extreme case, all subsets comprising ($m+1$) reactions form an EFM. Thus, with all reversible EFMs counted only once, S_{max} reads as follows:

$$S \le S_{\max} = \binom{q}{m+1} = \frac{q!}{(q-m-1)!(m+1)!}$$

$$= \frac{q \cdot (q-1) \cdot (q-2) \ldots (q-m)}{(m+1)!}$$

Fortunately, at least in biochemical networks, (2) seems to be a very pessimistic estimation. S_{max} is much reduced by irreversible reactions and by modes comprising only a small number of participating reactions. If an EFM can be found which utilises only w reactions ($w<m+1$) then S_{max} is reduced by the value

$$\binom{q-w}{m+1-w}.$$

The latter expression comprises all subsets with ($m+1$) columns (reactions) of N including the utilised w reactions. These subsets surely can not be elementary anymore. Enzyme subsets (reactions always operating together in the same rate ratio can also drastically reduce the upper bound because all reactions of an enzyme subset can be considered as one lumped reaction, thus, lowering the number q of reactions. If a reaction is added to a network then S_{max} increases by a factor of $(q+1)/(q-m)$. Certainly, the true S will never be smaller since the set of modes calculated without the inserted reaction is still valid. In particular, this means that analysing the EFMs of a subnetwork is still relevant for the overall network.

Inserting a new metabolite changes S_{max} by a factor of $(q-m-1)/(m+2)$. Thus, only if $m<(q-3)/2$, S_{max} increases, if $m = (q-3)/2$ it remains constant and otherwise it decreases. However, the upper bound S_{max} can decrease whereas the true number S increases. Especially highly connected metabolites increase S, because they can be balanced in many different ways by the use of different reactions. Thus, the relatively low connectivity number of most metabolites (cf. small world topology in metabolic networks is one reason for the observation that S_{max} is not reached in metabolic networks.

It is indeed difficult to estimate the number of EFMs occurring in a network more exactly from the number of participating metabolites and reactions alone. The number of modes even for networks having the same number of metabolites and reactions can differ by magnitudes. This fact emphasises the inherent structural information reflected by the elementary modes. Better estimations might be given by considering also the reversibility of the reactions or the connectivity of metabolites but require extensive theoretical work in the future.

EFMs in the Catabolic Network of Escherichia coli

In a case study, researchers calculated the EFMs in the central metabolism in Escherichia coli. The catabolic part was modelled in detail, whereas for the anabolic part predominantly lumped reactions were considered. Altogether, the network contains 89 metabolites and 110 reactions. The calculations were performed in the FluxAnalyser, a package for MATLAB (R) providing a user-friendly environment facilitating stoichiometric analyses of metabolic networks, in particular computationally efficient pathway analysis, on interactive flux maps.

Table 1. Number of EFMs for different substrates utilised by the network of Escherichia coli.

	Glucose	Acetate	Glycerol	Succinate	all 4 substrates simultaneously
Participating reactions/ metabolites	106/89	105/89	106/89	105/89	110/89
upper bound S_{max}	$3.69*10^{18}$	$5.57*10^{17}$	$3.69*10^{18}$	$5.57*10^{17}$	$4.39*10^{21}$
reduced network: participating reactions/ metabolites	46/30	45/30	47/31	45/30	51/31
reduced network: upper bound S_{max}	$5.12*10^{11}$	$1.67*10^{11}$	$7.52*10^{11}$	$1.67*10^{11}$	$4.85*10^{13}$
computation time	1027 sec	3.8 sec	53.6 sec	16.4 sec	approx. 50 hours
number of EFMs (S)	27099	599	11332	4249	507632

Apparently, the original size of the network and, hence, the upper bound S_{max} can be much reduced by the algorithm's preprocessing steps (Table 1). The most important reduction is obtained by lumping the biosynthetic routes (e.g. amino acid synthesis) into one overall biomass synthesis 'reaction' since they represent here an enzyme subset. However, the upper bounds S_{max} for the reduced models still significantly overestimate the actually observed number of EFMs.

In general, the computation time correlates well with the square of the final number of EFMs. One reason for this relation is, that preliminary modes must be compared with each other to remove nonelementary or duplicate modes. The computational demand for these checks is of order $o(P^2)$ (P: number of preliminary modes). Thus, in the case with simultaneous utilisation of all substrates, the considerable increase of the final number of EFMs is accompanied by a more than linear (but not exponential) increase of the computational demand.

A comparison of the number of EFMs shows the different flexibility of the network for utilising the different substrates. Here, high flexibility means a high number of alternative routes increasing the fault tolerance if a certain path fails. Glucose can be classified as the substrate which can be utilised as most flexible. For example, only the set of EFMs for glucose contains pathways enabling fermentative growth. Altogether, about 45 times more nondecomposable routes are available for glucose than for acetate. Compared to glycerol, about 2.5 times more EFMs occur, although S_{max} is higher for glycerol. It is tempting to speculate that a higher variety of degradation routes is one criteria for the cell to prefer certain substrates as it is the case for glucose.

Pathway analysis by elementary modes in larger metabolic networks turned out to be a computational challenge due to the combinatorial explosion of nondecomposable routes. An interesting fact is that in metabolic networks the number of EFMs grows exponentially with increasing network size but seems always to be far below the bound. Small routes or cycles, irreversible reactions, enzyme subsets and the relatively low connectivity of most of the metabolites dampen the explosion of the number of modes.

Analysing the core catabolism of E.coli for growth on four different substrates demonstrate that pathway analysis is feasible also in networks having higher complexity. The set of EFMs reflects important properties of the network, such as flexibility and functionality. For example, glucose could be classified as the substrate which can be utilised as most flexible. However, in networks comprising the complete set of biochemical pathways, it seems still to be impossible or would at least require parallel algorithms to calculate all EFMs. Approaches proposed for facilitating pathway analysis in complex networks mostly aim at the decomposition of networks into smaller subnetworks. Alternatively, one may consider subnetworks where reactions are neglected, since the resulting EFMs would also be valid for the complete network and, thus, a suitable approximation of the complete set of pathways.

CP(GRAPH)

CP(Graph) features graph variables, node and arc variables, and set variables for nodes and arcs.

Type	Representation	Constraint	Constants	Variables
Integer	0; 1; 2;...		i_0; i_1; :::	I_0;I_1; :::
Node	0; 1; 2;...		n_0; n_1; :::	N_0;N_1; :::
Arc	(0; 1); (2; 4);...		a_0; a_1; :::	A_0;A_1; :::
Finite set	{0; 1; 2}; {3; 5}...		s_0; s_1; :::	S_0; S_1; :::
Finite set of nodes	{0; 1; 2}; {3; 5}...		sn_0; sn_1; :::	SN_0; SN_1; :::
Finite set of arcs	{(0; 3); (1; 2)};... (SN; SA)		sa_0; sa_1; :::	SA_0; SA_1; :::
Graph	SN a set of nodes SA a set of arcs	$SA \subseteq SN \times SN$	g_0; g_1; :::	G_0;G_1; :::
Weight functions	$N \cup A \rightarrow \mathbb{N}$		w_0;w_1,...	

Three kernel constraints surfice to express all MS-definable properties of graphs as constraints:

Arcs(G; SA) SA is the set of arcs of G.

Nodes(G; SN) SN is the set of nodes of G.

ArcNode(A;N_1;N_2) The arc variable A is an arc from node N_1 to node N_2. This relation does not take a graph variable into account as every arc and node has a unique identifier in the system. If A is determined, this constraint is a simple accessor to the tail and head of the arc A and similarly if both nodes are determined.

These constraints enable to express more complicated constraints such as Reachables (G;N; SN) which states SN must be the set of nodes reachable from N in G (the transitive closure of the adjacency relation in G) or Path(G;N_1;N_2) which holds if G is a path from node N_1 to node N_2. While these constraints can be expressed using kernel constraints, they are more efficient when implemented using dedicated global propagators (either for their consistency level or algorithmic complexity).

CP(Graph) enables to express constrained subgraph extraction problems such as the TSP or the equicut problem:

— Finding the TSP in graph g with weights w: minimise Weight(G;w) s.t. SubGraph(G; g) ^ Cycle(G) ^ Nodes(G) = Nodes(g)

— Graph partitioning problem: equicut of a graph g of even order: minimise #(Arcs(g) n (Arcs(G1) [Arcs(G2))) s.t. SubGraph(G_1; g)^SubGraph(G_2; g)^Nodes(G_1)∪[Nodes(G_2) = Nodes(g)^ #Nodes(G_1) = #Nodes(G_2) =1/2 #Nodes(g)

Constraint programming was used to solve constrained path finding problems in using a finite domain model (successor variables). Path variables were introduced in to solve constrained path problems as part of a network design problem. A cost-based filtering technique for constrained shortest path problems was described in and a global path constraint was presented in.

Graphs also play an important role in constraint programming in the design of propagators for global constraints: graph algorithms are used and global constraints were modeled as networks of similar constraints. While queries about reachability can be handled by both systems, it seems that some queries such as optimisation problems can be expressed using CP(Graph) but not with CTL logic. As it is, CP(Graph) allows constrained subgraph extraction.

Metabolic Network Analysis

The general kind of analysis researchers wish to perform with CP(Graph) is pathway discovery by constrained subgraph extraction. One potential application of this type of analysis lies in assisted explanation of DNA chip experiments. In such experiments, the behaviour of a sane cell and a mutant are compared in a given context (the substrate on which they are living or more generally their environment).

This comparison is done at different times by extracting and amplifying the expressed RNA in the nucleus of the cells (this kills the cell). This RNA is then put on a DNA chip: an array of representative sequences of bases for a set of genes. The RNA binds to the chip in the locations which are specific to it. That array is then scanned to see the level of expression of each RNA strand. That RNA encodes for given enzymes which catalyse given reactions.

Hence, the level of RNA can be translated into the information of which reactions were active in the cell at the time its RNA was extracted. Given this set of reactions, biologists would like to know which processes where active in the cell. If a CSP allows to recover known processes from sets of reactions, it could be adequate to discover the real processes given other sets of reactions. Hence, such a CSP could approximate the real processes at work in a cell from DNA chip results. These computational results could then be used to further guide other concrete experiments which are more expensive.

This experiments focus on linear pathways by doing constrained shortest path finding. Experiments comprise increasingly better characterisations of the CSP (ie. more constraints which increase the rate of correct recovery) and a formulation of a CSP for pathways which contains branchings or cycles.

Prototype of CP(Graph) Implemented in Oz/Mozart and Gecode

A prototype of CP(Graph) in the Oz/Mozart constraint

programming framework. A set of nodes and a set of arcs are used to implement each graph variable.We also implemented this prototype over the Gecode generic constraint development environment. In these prototypes, they implemented, among others, the global path propagator of.

Constrained Shortest Path Finding

As about half of the known pathways are simple paths, one type of experiment consists in trying to find these pathways by using constrained path finding in a directed graph (knowing a few nodes of the path). In, experiments were done first with a dedicated shortest path finding algorithm. Then some nodes (the pool metabolites, molecules like ATP or H_2O which are ubiquitous and take part in many reactions) were removed from the graph and the results compared with the previous ones.

Some pathways, such as glycolysis, however use some of these metabolites as intermediates. In order to decrease the likelihood of selecting these nodes while still allowing to select them, all nodes were assigned a weight proportional to their degree. As pool metabolites have a very high degree, they are much less likely to be selected in the shortest paths. This experiment consists in redoing the former experiment with an additional constraint of mutual exclusion for certain pairs of reactions. These pairs are reverse reactions (the reaction from substrates to products and the one from products to substrates).

Most of the time, these reactions are observed in a single direction in each species. Hence they wish to exclude paths containing both in this experiment. Such additional constraints like mutual exclusion are not always easily integrated in dedicated algorithms. In CP(Graph) it just consists in posting a few additional constraints. If n_1; :::; n_m are the included reactions and (r_{i1}; r_{i2}); $0 < i \leq t$ the mutually exclusive pairs of nodes, the programme looks like: miminise Weight(G;w) s.t.

$$\text{SubGraph}(G;\ g) \wedge \text{Path}(G;\ n_1;\ n_m) \wedge \forall 0 < i \leq m : n_i \in \text{Nodes}(G) \wedge\ \forall i \in [0;\ t]\ :\ r_{i1} = \notin \text{Nodes}(G) \vee\ r_{i2} \notin \text{Nodes}(G)$$

In their experimental setting researchers first extract a subgraph of the original metabolic bipartite digraph by incrementally growing a fringe starting by the included nodes. Then, given a subset of the reactions of a reference pathway, they try to find the shortest constrained path in that subgraph. The first process of extraction of a subgraph of interest is done for efficiency reasons as the original graph is too big to be handled by the CSP (it contains around 16.000 nodes).

The running time increases greatly with the size of the graphs. The programme can however be stated in a few lines and first results obtained the same day the experiment is designed. The limitation on the input graph size does not guarantee to get the optimal shortest path in the original graph. This should however not be a major problem as biologists are most of the time interested in a particular portion of the metabolic graph.

The rapidity of expression and resolution of such a NP(Hard) problem outweights this size limitation. The first is being able to cope with bigger graphs. They could design more efficient heuristics for labelling. The use of a cost-based filtering method could prune the size of the graph given an upper bound of the cost. Such an upperbound is available as soon as a first solution is found. Another solution would be to use an a-priori upper bound of the cost which would need to be increased or removed if no solution is found.

The second aspect of this experiment consists in finding which additional constraints are needed to recover known pathways as it was shown in that non-constrained shortest paths are not able to recover all of them. They are currently working on a extention of this approach to discovering pathways containing branchings or cycles. A first formulation they wish to test is the following: find the smallest graph containing all the seeds such that there is a seed from which all other nodes are reachable. Given sn_s a set of nodes (seeds), minimise Weight(G;w) subject to:

$$N \notin sn_s; sn_s \subseteq \text{Nodes}(G) \wedge \text{Reachable}(G;N;\text{Nodes}(G))$$

7

Topological Features of Metabolic Networks

There has recently been an upsurge of interest in the socalled scale-free networks, where the vertex order, or degree of connectivity per node k, follows a power-law distribution

$$P(k) \sim k^{-\gamma} \quad (1)$$

for k> ⟨k⟩. This is in contrast to the exponential suppression expected from randomly wired networks. For a number of real-world networks like social networks, power grids, citation networks, the world wide web, and metabolic networks, the scale-free behaviour of Eq. (1) has been observed with exponents g in the range 2-4, at least within an appropriate range of orders. The nature of all examples above is that they originate from a growing process. Hence, it appears natural to develop a model of growing network, which was indeed done in.

The degree of connectivity is a measure of local properties of the networks. Less attention has been payed to more global properties for the above examples except for general features like network diameters. Global measures could be important discriminants of network properties, in particular, when it comes to narrow on potential underlying models.

The citation and metabolic networks are quite low dimensional with d_S around 3, whereas for the growing network model d_S is approximately 7.5. The latter turns out to be more in parity with what characterises random and small-world networks. As a consistency check, the dimension of regular lattice networks is also determined, which comes out as expected. One concludes that the spectral dimension offers a powerful and additional measure to γ in Eq. (1) when it comes to characterising network topologies.

Furthermore, the diffusion process underlying the extraction of dS hints upon differences in the citation and metabolic network topologies; the former is treelike whereas the latter is rich in loop structures.

SPECTRAL DIMENSION

To probe the topological properties of an interaction network, which is of more global nature than the degree of connectivity, is based upon the diffusion of a test particle in the metric space defined by the graph. In a continuous space with a fixed smooth metric, the diffusion equation has the form

$$\frac{\partial}{\partial 1} K_g(\in, \in_0, t) = \Delta_g K_g(\in, \in_0, t) \qquad (2)$$

Here t is the diffusion time, Δ_g is the Laplace operator in the metric g, and $Kg(\in, \in_0, t)$ is the probability density to diffuse from $\in_0$ to $\in$ in time t. For small t it is well known, that the average return probability has the following asymptotic expansion

$$K_g (0,0,t) \sim t^{-d_S/2} \qquad (3)$$

with

$$K_g (\in, 0, 0) = \delta (\in) \qquad (4)$$

and where the power $t^{-dS/2}$ reflects the dimension of the network.

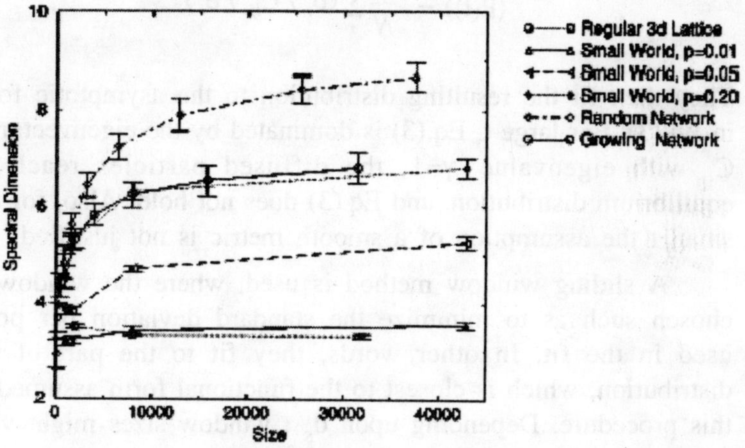

Figure 1. Spectral dimensions dS as functions of size for synthetic networks.

In the spirit of these equations they extract the spectral dimension of the geometry defined by the interaction networks considered in this work. To this end, they use the transition matrix

$$C_{ij} = \frac{J_{ij}}{k_i} \tag{5}$$

with

$$J_{ij} = \begin{cases} 1 \text{ if } i \text{ and } j \text{ are neighbors} \\ 0 \text{ otherwise} \end{cases} \tag{6}$$

as the discrete version of the Laplace operator Δ_g. The number k_i, counts the number of links connected to vertex i—the vertex order. In a simulation of the diffusion process defined by the transition matrix, the probability to return, P, after t steps to the origin δ_i is then measured. More precisely, they choose a random subset $\{\delta_i\}$%,i=1…N of vertices and extract

$$\langle P(t) \rangle = \frac{I}{N}\sum_{t}^{N}\langle \delta_i \,/\, C_{ij}^{T} \,/\, \delta_i \rangle \quad (7)$$

They then fit the resulting distribution to the asymptotic form in Eq.(3). For large t, Eq.(3) is dominated by the eigenvector of C_{ij} with eigenvalue $\gamma=1$; the diffused particles reach an equilibrium distribution, and Eq.(3) does not hold. Also, for too small t the assumption of a smooth metric is not justified.

A sliding window method is used, where the window is chosen such as to minimize the standard deviation per point used in the fit. In other words, they fit to the part of the distribution, which is closest to the functional form assumed in this procedure. Depending upon d_S, window sizes might vary from a few time points to the entire range.

SYNTHETIC NETWORKS

Next extract d_s for a few different types of synthetic networks with sizes up to around 30 000 nodes in order to compare with typical sizes of citation networks. In addition, for all networks the average geodesic distance (diameter) R is computed.

Lattice Networks

Here use a simple square lattices with dimension d=3. For these it is only possible to return to the origin after an even number of steps. Therefore P(t)=0 for odd t. In terms of the transfer matrix eigenvectors this means that the symmetry of the square lattice leads to an eigenvector with λ exactly –1.

The corresponding eigenstate does not decay for large t and yields destructive interference with λ =1 eigenstate after an odd number of steps. When extracting the spectral dimension dS from the data they therefore omit the odd time points. As can be seen from Fig. 1, d_s is slightly above the expected value 3, the dimension of the lattice. Researchers have repeated this experiment for a regular lattice including diagonals, hence allowing for closed loops with an odd number of steps.

Although local properties of the graph are considerably changed since the number of links emerging from each vertex is increased from 2d to 2d+d(d–1), they still extract the same value for d_s within errors. This is a further justification of omitting the data at odd time points for the square lattice and emphasises that the spectral dimension is sensitive to global geometric aspects rather than to local details.

Small-world Networks

They generate small-world networks from the lattice network described above by rewiring the edges in the regular lattice to a randomly chosen vertex with probability P = 0.01, 0.05, and 0.2 respectively. This generates three sets of models for each system size. Not surprisingly, these networks end up with spectral dimensions (Fig. 1) in between the regular lattice and the homogeneous random networks described next.

Random Networks

Homogeneous random networks are generated as directed networks, where each vertex has two inputs from other vertices chosen at random. In the analysis the graph is considered undirected by ignoring the orientation on the edges connecting the vertices. Furthermore, the analysis is focused on the largest cluster; the largest connected part of the resulting graph.

The spectral dimension $d_s \approx 8.5$ (Fig.1) is relatively large. This agrees qualitatively with the observation that a characteristic length scale, e.g., average distance between vertices along links, grows slowly with the volume. It also signals a large dimensionality for this type of graph. It is interesting to note that again a distinct even-odd disparity emerges; the amplitude after an odd number of steps is almost zero.

This indicates that the random network is dominated by a treelike structure, the number of loops (at least of odd length) is negligibly small compared to the amplitude obtained from backtracking the same path.

Growing Network Model

In this model one at each time step t adds a node, which connects to m existing nodes (i) with a probability Π given by

$$\Pi_i = \frac{k_i^{\beta}}{\sum_j k_j^{\beta}} \tag{8}$$

where k_i is the connectivity of node i and β is a parameter. In the β=1 case was investigated both analytically in the t→∞ case and numerically for finite t. With the approximations involved one arrived at $\gamma = 3$ in the analytic approach and $\gamma = 2.9 \pm 0.1$ from simulations. For these networks, R is approximately 6 for large system sizes.

Citation Network

In a citation network the links and nodes correspond to the citations and the publications (citing and cited), respectively.

By considering graphs generated by publications in a certain year or time span and the cited papers (regardless of publication year) they obtain a whole set of graphs. In these not fully connected graphs they focus on the largest connected cluster. When computing the connectivity distribution of the nodes, they confirm the power-law distribution (Eq.1) already observed in citation networks with $\gamma = 2.7$.

In Fig. 2 they plot the probability to return to the origin after t time steps for a few typical networks. It is interesting to note that, within errors, the return probability for the 76 and 79 networks are in very good agreement indicating universal topological properties. This holds true also when comparing all of the other one-year networks with the exception of the 75 and the 87 networks (not shown), which cannot be fitted to Eq. (3). In these years the connectivity does not probably generate a homogeneous metric, because the geometry is dominated by two or more large clusters, which are interconnected by only a few links.

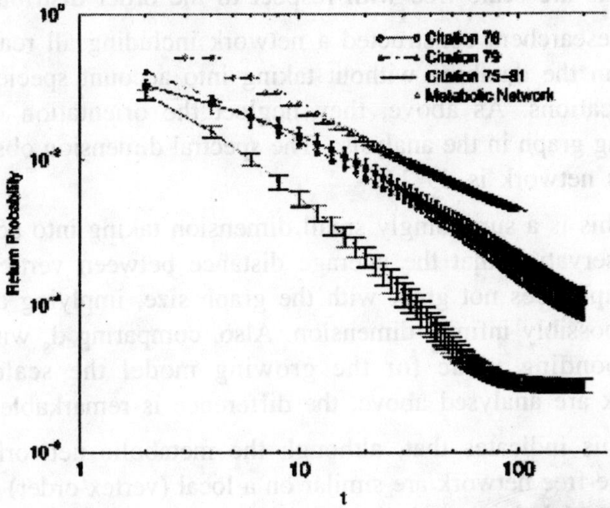

Figure 2. The return probability (P) as a function of time steps t for citation networks.

For the networks composed by several years they again observe a remarkable universality when comparing different time periods. However, the distribution observed is considerably different from the one-year distributions. The reason presumably is a maturation of the citation networks leading to a tighter interconnection of the central cluster. In turn they observe larger dimension as can be seen from the steeper slope of the corresponding graph in Fig. 2. These issues for the citation networks have been subject to further investigations.

Metabolic Network

Another real-world interaction network is the metabolic network found in living cells. In these networks, substrates are treated as vertices, while chemical reactions connecting substrates and educts are treated as directed links. Recently it has been demonstrated that for this type of networks the connectivity distribution obeys a power-law behaviour (Eq. (1)); these networks are scale free with respect to the order distribution.

Researchers constructed a network including all reactions found in the database without taking into account species and cell locations. As above, they neglect the orientation of the resulting graph in the analysis. The spectral dimension observed for this network is $d_s \approx 2.8$.

This is a surprisingly small dimension taking into account the observation that the average distance between vertices on this graph does not grow with the graph size, implying a very large, possibly infinite dimension. Also, comparing d_s with the corresponding value for the growing model the scale-free network are analysed above, the difference is remarkable.

This indicates that, although the metabolic network and the scale-free network are similar on a local (vertex order) scale, the more global topological properties are very different. Another important difference arises from the observation that the return probability for odd path lengths do not vanish. This means that the metabolic networks consist of a very large number of closed loops. The graph has, in contrast to the random, growing model and citation networks, not a dominantly treelike structure.

8

Structure of Expanding Metabolic Networks

Traditionally, mathematical modelling of metabolic networks aims at simulating stationary states or time courses for the system variables with the system parameters used as input quantities. Recently, structural analyses gained much interest since in contrast to enzyme kinetic parameters the stoichiometric properties are often much better known . The development of metabolic databases such as KEGG (Kyoto Encyclopedia of Genes and Genomes) allows for studying the structure of metabolism on a large scale.

In previous work some researchers performed structural analyses on subnetworks by applying evolutionary optimisation using a limited set of generic reactions. They show that the stoichiometry of the central energy metabolism, in particular the ordering of ATP and NADH producing and consuming reactions, can be explained under the premise that contemporary networks ensure a high ATP production rate. Furthermore they introduced the concept of stoichiometric robustness characterising the effect of structural changes on the function of metabolic networks.

Then they reconstruct extant metabolic systems on a much larger scale by taking into account as many specific reactions as possible. They perform this reconstruction by simulating evolutionary processes starting from a small set of reactions and compounds. During this procedure new reactions are repeatedly attached according to certain rules leading to networks of increasing size. Each individual process they call network expansion. They show that the mutual dependencies of reactions by their substrates and products allow for conclusions on the temporal order of the emergence of enzymatic reactions and subpathways.

Such studies will help to elucidate the principles under which network structures evolved towards increasing complexity. In this work they develop methods for this kind of investigation which are tested first on the limited set of reactions of the glycolytic pathway. Application to the large set leads to the definition of new structural characteristics such as the scope of compounds.

Any metabolic network depends on substrates which have to be provided by the environment. Through a series of biochemical reactions these substrates get converted into end products which are released into the environment. Such substrates and products are called external compounds of the given network. Under steady-state conditions, the production and consumption of all other (internal) metabolites is balanced resulting in time independent concentrations. Let $C = (X; S)^T$ denote the vector of the concentrations of the metabolites such that X comprises all internal and S all external compounds. The dynamic behaviour of the network is governed by a set of differential equations for the metabolite concentrations having the form

$$\frac{dC}{dt} = N \cdot V \qquad (1)$$

where N is the stoichiometric matrix which can be decomposed into two submatrices N_X and N_S for the internal and external compounds, respectively, and V is the vector of the rates of the biochemical reactions. In principle, all reactions are reversible allowing for positive and negative values of V. Under physiological conditions, however, many reactions proceed only in one direction leading to sign restrictions for the corresponding components of V. In the following, they consider all reactions tǒ be reversible. The steady-state of the system is determined by the equation

$$N_x \cdot V = 0 \tag{2}$$

In the framework of stoichiometric analysis, Eq. (2) is treated as a linear equation system for feasible uxes V. In general, the solution of the system is not unique with the general solution being a superposition of a set of fundamental solutions. They distinguish between two different types of solutions, one where the uxes result in a net conversion of external metabolites, $N_x \cdot V \neq 0$, and one where the uxes represent only internal cycles, $N_s \cdot V = 0$.

In the present work they investigate expanding network structures leading to sequences of stoichiometric matrices of increasing size. This analysis is performed in view of evolution where metabolic networks have acquired new metabolic capabilities in form of additional enzymatic reactions. On shorter time scales, changes in network size could be mediated by a gene transfer as can be observed between bacteria. For deriving rules for network expansion, they assume that a newly acquired biochemical reaction may be especially useful to an organism if it makes it less dependent from external resources or provides it with new resources. This can be fulfilled by reactions allowing to treat one or more compounds to be internal which were external before the acquisition.

If the acquired reaction depends on metabolites not present in the network before, these compounds must be considered external. However, for some chemicals which were always highly abundant in the environment there was no need for an organism to balance their production and consumption. In their analysis they consider such compounds to remain external during the whole process of network expansion. They investigate several scenarios with different sets of these highly abundant compounds.

The computational analysis of network expansion is performed by an algorithm which can be summarised as follows:

1. The process starts with a set of one or more biochemical compounds acting as a seed of the expanding network.

2. From the set of all available reactions a set of candidate reactions is determined. Candidate reactions are defined by the property that they can in principle be attached to the current network. Candidate reactions must fulfil one of the two expansion conditions: At least one of its compounds must already be present in the current network (weak expansion condition $-C_1$); all of its substrates or all of its products must already be present (strong expansion conditio $- C_2$).

3. From the set of candidate reactions, one or more are attached to the current network according to specific selection rules. In this work, the selection rules utilise Eqs. 2 and 1, for example for deciding which reactions perform net conversions under steady state conditions. Steps 2 and 3 are repeated continuously until no candidate reaction can be found. Each loop leads to a new network generation.

Expansion on the Set of Glycolytic Reactions

This section describes the network expansion on a limited set of reactions. Candidate reactions are chosen from the thirteen reactions of the glycolytic pathway (Figure 1). Since this pathway supplies energy in form of ATP, consumption of ATP by other processes (ATPases) is also taken into account (the reaction placed on top of the first glycolytic reaction at the left end of Figure 1).

They analyse network expansion on this set of reactions under the following conditions. First, they assume that the network in each stage depends on as few as possible external metabolites. Second, they assume that the network produces as many as possible internal metabolites since these compounds may serve as building blocks for other metabolic processes. Third, they consider it disadvantageous for a network to possess reactions that are not required for conversions under steady-state conditions. These reactions are characterised by the fact that the corresponding entries in all solution vectors V of Eq. (2) vanish.

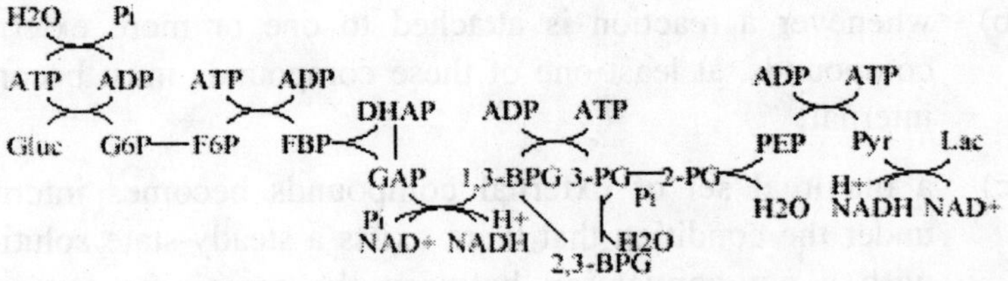

Figure 1. Base set of reactions comprising the glycolytic pathway. Abbreviations for the compounds are: Gluc - glucose, G6P - glucose-6-phosphate, F6P - fructose-6-phosphate, FBP - fructose-1,6-bisphosphate, DHAP - dihydroxyacetone phosphate, GAP - glyceraldehyde-3-phosphate, 1,3-BPG - 1,3-bisphosphoglycerate, 2,3-BPG - 2,3-bisphosphoglycerate, 3-PG - 3-phosphoglycerate, 2-PG - 2-phosphoglycerate, PEP - phosphoenolpyruvate, Pyr - pyruvate, Lac - lactate.

The selection rule can be formalised as follows: From the candidate reactions, those reactions are chosen maximising a selection value for which they use the expression

$$f = \alpha x - \beta s - \gamma z; \qquad (3)$$

where x and s are the numbers of internal and external metabolites, respectively, and z is the number of reactions that are not required in steady-state and α, β, γ and are weight factors. The presented simulations have been performed for $\alpha = \beta$ and $\gamma < \beta$. The first choice ensures that the value of f remains unchanged if the acquisition of a new reaction involves the incorporation of a new external compound into the network and simultaneously an external compound becomes internal.

The second choice entails that an additional reaction which is not required for conversions is less disadvantageous in terms of the selection value *f* than an additional external compound. Additional criteria to be fulfilled during the process are:

a) an internal compound cannot become external in later generations;

b) whenever a reaction is attached to one or more external compounds, at least one of these compounds must become internal;

c) a maximal set of external compounds becomes internal under the condition that there exists a steady-state solution with a net conversion between the remaining external metabolites ($N_S - V \neq 0$).

In the set of glycolytic reactions there are certain compounds whose production and consumption are strictly coupled. These are ATP and ADP as well as NAD^+, NADH and H^+. Therefore, the stoichiometric matrix is reduced by considering only one compound from each group.

Model Scenarios

Using the algorithm the analyse three scenarios. All of them are based on the weak expansion condition C_1, using glucose as the only initial compound and considering water as highly abundant. The three scenarios differ in the choice of compounds which, in addition to water, are forced to be external and cannot become internal during the expansion. These compounds are selected from the metabolites ATP, NAD^+ and inorganic phosphate.

ATP, NAD^+ and inorganic phosphate as forced external compounds

In the first step of the algorithm, there is only one candidate reaction (hexokinase – HK) which can be attached to glucose. In this stage the networks contains three new compounds, all of which are external. Two of them (ATP and ADP) are forced to be external whereas G6P can become internal later (Figure 2). Like in the first step, for each of the following three steps there exists only one candidate reaction. This leads to the generation 4 which includes the reactions HK, glucosephosphate isomerase (PGI), phosphofructokinase (PFK) and aldolase (Ald).

In the fifth step there are two candidate reactions, triose phosphate isomerase (TIM) and glyceraldehyde phosphate dehydrogenase (GAPDH). Incorporation of TIM leads to generation 5 whereas incorporation of GAPDH leads to a network depicted by 5* in Figure 2.

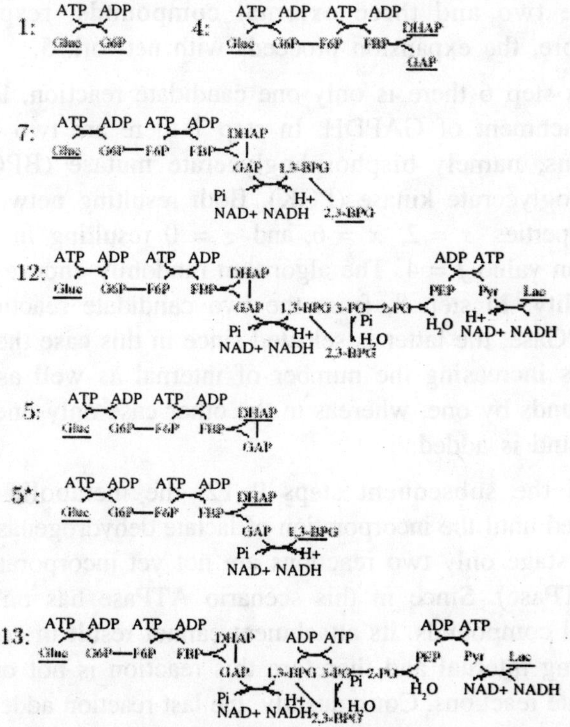

Figure 2. Expansion process on the limited set of the glycolytic reactions (case 1).

Figure 2 assumed that water, ATP, NAD$^+$, and inorganic phosphate are highly abundant in the environment and forced to be external. These compounds are marked in grey. ADP, NADH, and H$^+$ are strictly coupled to other compounds and also do not contribute to the selection value f and are therefore marked in grey as well. Compounds which are external in one generation but may become internal in a later generation are underlined. The calculations were performed with the parameters $\alpha = \beta = 1$ and $\gamma = \beta/2$.

These two possibilities differ in their relations between the numbers of external and internal compounds. Whereas both networks contain four internal compounds, networks 5 and 5* include two and three external compounds, respectively. Therefore, the expansion proceeds with network 5.

In step 6 there is only one candidate reaction, leading to the attachment of GAPDH. In step 7 there are two candidate reactions, namely bisphosphoglycerate mutase (BPGM) and phosphoglycerate kinase (PGK). Both resulting networks have the properties $s = 2$, $x = 6$, and $z = 0$ resulting in the same selection value $f = 4$. The algorithm randomly chooses the first possibility. In step 8, from the two candidate reactions PGK and BPGase, the latter is selected since in this case the network expands increasing the number of internal as well as external compounds by one, whereas in the other case only one external compound is added.

In the subsequent steps 9–12, the metabolic chain is expanded until the incorporation of lactate dehydrogenase (LDH). In this stage only two reactions are not yet incorporated (PGK and ATPase). Since in this scenario ATPase has only forced external compounds, its attachment cannot result in compounds becoming internal and therefore this reaction is not one of the candidate reactions. Consequently, the last reaction added is PGK leading to generation 13 representing the full glycolytic pathway.

Inorganic phosphate as forced external compound

As in case 1, the first step represents the incorporation of HK.

Since in this scenario ATP can become an internal compound, the expansion continues in a different fashion compared to case 1. In fact, there are five candidate reactions in step 2, four of them proceed with ATP turnover and are therefore suited to make ATP an internal substrate when combined HK. The fifth candidate reaction is PGI. The algorithm selects ATPase since no new external compounds are incorporated. From this step the expansion proceeds in the same way as in case 1.

No additional forced external compound

In this scenario in principle all compounds except water can become internal during the expansion. The first six steps proceed in the same way as in case 2 resulting in the network 6 in Figure 3. This networks converts glucose into two molecules DHAP and consumes two molecules of ATP which are produced by ATPase working in the opposite direction by the expense of inorganic phosphate which in this stage is external.

The expansion proceeds by prolongation of the chain from 3PG to pyruvate. This leads in step 10 to a pathway consuming besides glucose also 2,3-BPG with a net consumption of one ATP. ATPase still works in the opposite direction where the rate is halved compared to network 6. In steps 11 and 12 the two sugar converting chains are connected by incorporation of BPGM and GAPDH. This network transforms glucose into pyruvate with no net consumption of ATP. Therefore, ATPase occurs with a zero rate resulting in a value $z = 1$. In step 13 LDH is attached allowing for the internalisation of NAD^+ and pyruvate which were external in step 12.

Only with the incorporation of PGK in the last step, the glycolytic pathway achieves the ability to produce ATP. Consequently, ATPase now operates in the forward direction. They chosen the well known glycolytic pathway to demonstrate in detail the procedure of network expansion. Despite its simplicity the expansion may proceed in many modes differing in the order of attachment of the reactions.

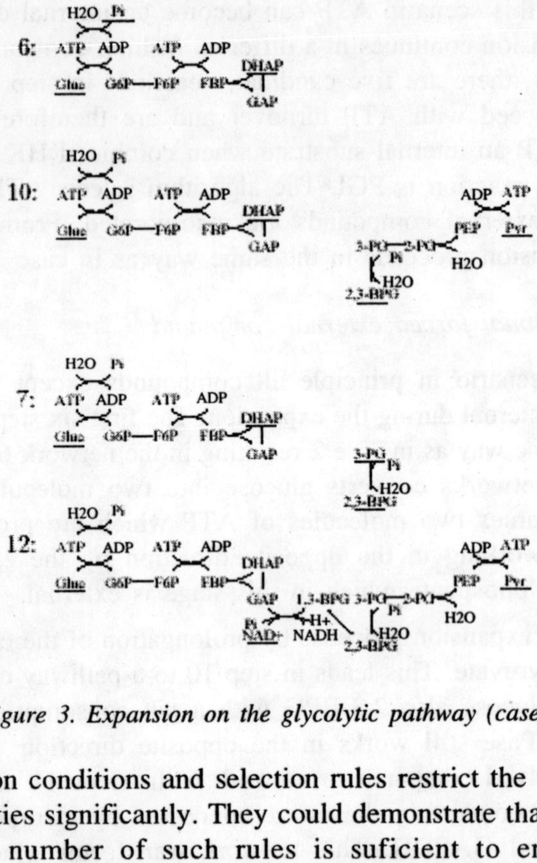

Figure 3. Expansion on the glycolytic pathway (case 3).

Expansion conditions and selection rules restrict the number of possibilities significantly. They could demonstrate that applying a small number of such rules is suficient to ensure that intermediary generations do not represent arbitrary collections of reactions but rather networks with preliminary capabilities. Interestingly, the same criteria did not lead to a premature end but the expansion evolved until all reactions of the glycolytic pathway were incorporated.

Network Expansion on a Large Scale

They now apply the expansion methods on a set of 5161 reactions and 4450 compounds as provided by the KEGG database.

Strong Expansion Condition without Selection Rules

First, they investigate the expansion on this global set of reactions according to the following rules:

a) Several randomly chosen metabolites serve as initial compounds;

b) the expansion condition C_2 is used which is more restricted than condition C_1;

c) all candidate reactions are attached when proceeding from one generation to the next.

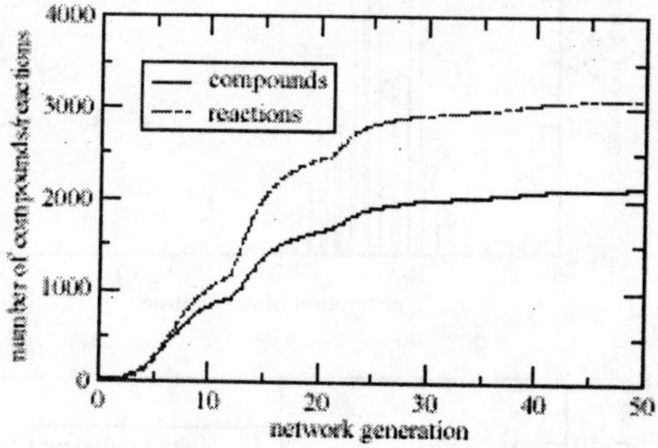

Figure 4. Evolution of the number of compounds and reactions during network expansion.

Figure 4 shows the development of the total numbers of compounds and reactions during one particular expansion. The curve is characterised by a strong initial increase and saturation in later generations. Interestingly, the algorithm stops after 50 generations resulting in a network of about 2100 compounds which is considerably smaller than the global network. This behaviour is due to the fact that at this stage no candidate reaction can be found fulfilling the strong expansion condition C_2. Similar results were obtained for other sets of initial compounds leading to final networks of different sizes.

The strong initial increase of the network size results from the fact that in this phase with each additional reaction a large number of new compounds are recruited which in turn increase the number of candidate reactions for the next generation. Another interesting feature is the tendency of the curve to level off in intermediate stages followed by another strong increase. They explain this by the existence of "bottlenecks" in the global network where only few reactions link highly connected domains.

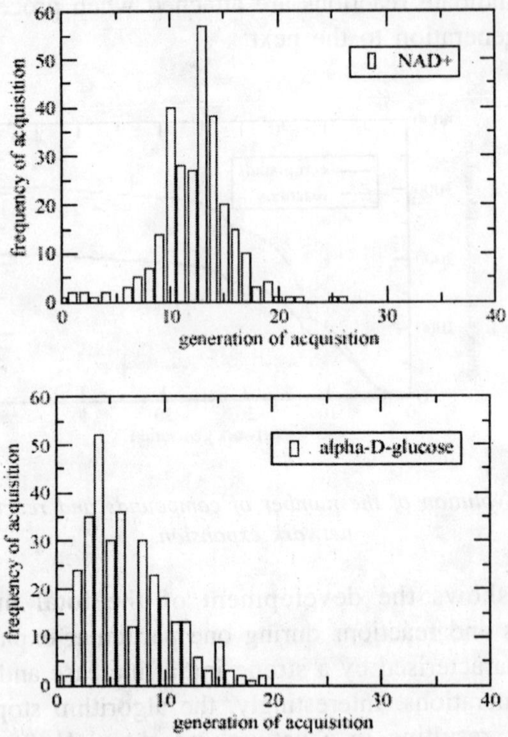

Figure 5. Histograms of the network generations at which the compounds glucose and NAD$^+$ are acquired during the expansion.

An intriguing question is in which generation a certain compound is acquired. Naturally, the particular value is highly dependent on the choice of initial compounds. They have therefore run a large number of simulations, each starting from a random initial set of 20 compounds. Figure 5 represents the results for the two compounds NAD+ and α-D-glucose. The diagram shows the frequency of acquisition in a certain generation. It is observed that glucose is generally acquired earlier during the expansion than NAD+. The mean generations of acquisition for NAD+ and glucose are 12.4 and 6.8, the standard deviations are 3.3 and 3.8, respectively. Figure 6 shows the number of compounds having a certain mean generation of acquisition.

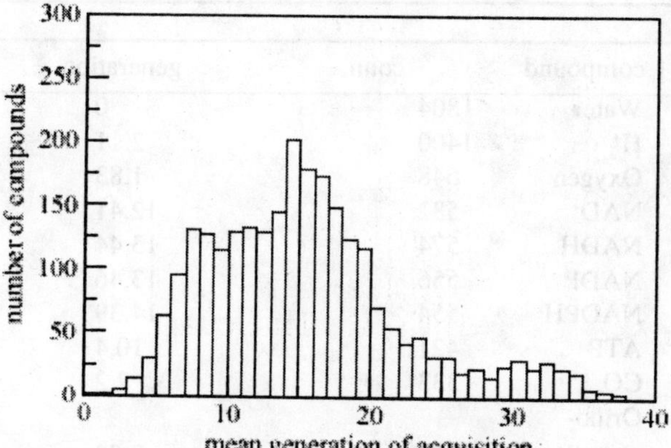

Figure 6. Histogram of mean generations of acquisition.

In table 1, top part, are listed those ten compounds having the lowest mean generation of acquisition. For each compound the number of reactions is listed in which it participates as substrate or product (connectivity). In the right part they list those ten compounds having the highest connectivity.

Table 1: The ten compounds with the lowest mean generation of acquisition (top) and the ten compounds with the highest connectivity (bottom).

compound	generation	conn.
Water	0	1804
Oxygen	1,83	648

H^+	2,11	1400
CO_2	2,2	333
H_2O_2	2,33	147
NH_3	2,75	274
Ortho-phosphate	3,03	329
O_2^-	3,25	9
Carbonic acid	3,42	3
Pyruvate	3,56	146

compound	conn.	generation
Water	1804	0
H^+	1400	2,11
Oxygen	648	1,83
NAD^+	582	12,41
NADH	574	13,44
$NADP^+$	556	13,46
NADPH	554	14,39
ATP	421	10,4
CO_2	333	2,2
Ortho-phosphate	329	3,03

It can be seen that five compounds appear on both sides meaning that compounds with a high connectivity tend to be acquired early in the expansion. Interestingly, there is no one to one correspondence between these two tables, an observation which will be described below.

Scope of Compounds

Network expansion can also start with a single initial compound (A). Again they apply the strong expansion condition C_2 resulting in a final network containing besides A all possible compounds which can be synthesised from A. This set of compounds they denote by $\Sigma(A)$ and call it the scope of A. If a compound B lies in the scope of A, then the scope of B is a subset of the scope of A:

$$B \in \Sigma(A) \Rightarrow \Sigma(B) \subseteq \Sigma(A) \tag{4}$$

There exist compounds which can be interconverted. For two such compounds A and B, the following implication holds true due to relation (4):

$$B \in \Sigma(A) and A \in \Sigma(B) \Leftrightarrow \Sigma(A) = \Sigma(B) \tag{5}$$

Figure 7 shows a histogram of the scope sizes. It can be observed that most scopes are of a size smaller than 500. There are only a few scopes of very large size. Not all scope sizes occur leading to gaps in the histogram. Scopes exist with almost all sizes below 50.

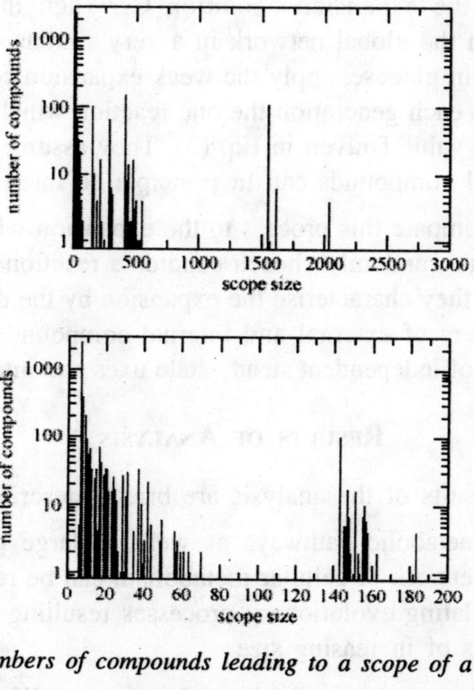

Figure 7. Numbers of compounds leading to a scope of a certain size.

Of special interest are the scopes of large sizes. Closer inspection shows that the largest set of 2042 compounds represents one scope that can be reached from four initial compounds which are 3'-phosphoadenosine 5'-phosphosulfate (PAPS), adenosine 5'-phosphosulfate (APS), dephospho-CoA, and UDP-6-sulfoquinovose. From the property of scopes expressed by Eq. (5) it follows that these four compounds are interconvertible. Similarly, the set of 1545 compounds represents a single scope which can be reached by 105 initial compounds. Among these are central cofactors such as ATP, GTP, NADH, NADPH as well as the compound UDPglucose. Further investigation reveals that the latter scope is a subset of the largest scope. This means, for example, that ATP can be synthesised from PAPS whereas the opposite process is not possible.

So far they expanded the global network by adding at the transition from one generation to the next all candidate reactions allowed by the expansion condition C_2. Then they perform expansion on the global network in a very similar way. Again, they start from glucose, apply the weak expansion condition C_1, and attach in each generation the one reaction which maximises the selection value f given in Eq. (3). They assume that during expansion all compounds can in principle be internalised.

They compare this process to the expansion where in each generation one randomly chosen candidate reaction is attached. In each case they characterise the expansion by the development of the numbers of external and internal compounds as well as the numbers of independent steady state uxes and internal cycles.

Results of Analysis

The main results of the analysis are briefly described below.

— Single metabolic pathways as well as large parts of the global network of cellular metabolism can be reconstructed by simulating evolutionary processes resulting in series of networks of increasing size.

— Depending on the availability of resources, different modes of expansion lead to the glycolytic pathway.

— Expansion on the global network yields a ranking of the metabolites according to the mean generation in which they were acquired when starting from randomly selected initial metabolites. All compounds which are on average acquired early in the expansion are well known core components of the metabolism and moreover are rather simple in their chemical structure.

4) Compounds of a network may be characterised by their scope defining the set of substances that can be synthesised through the available reactions without further use of other resources. Compounds with a large scope are found to be composed of a rather large number of chemical groups.

— Expansion processes obeying selection rules proceed significantly different to random expansion. Selected expansion results in networks being able to perform relatively few interconversions but exibly via various different internal routes whereas random expansion results in networks which can perform many interconversions but are highly dependent on external compounds.

The compounds with a low mean generation of acquisition are not necessarily those participating in many reactions. Carbonic acid, for example, is acquired very early in the expansion but has an extremely low connectivity (Table 1). This can be understood by the fact that it can be produced from water and carbon dioxide both having a very high connectivity and are acquired very early. Moreover, metabolites with high connectivities are not necessarily acquired early. Examples are NAD^+ and other cofactors. This can be explained by the fact that the synthesis of these compounds require a large number of steps.

The present investigation was performed by considering all reactions as reversible. For some reactions it is well known that they proceed only in one direction under physiological conditions. This information could easily be taken into account by applying a sign restrictions for the corresponding elements of the feasible solution vectors of Eq. (2). They refrained from including this information since the results would be dependent on the concentrations of external metabolites which do not represent a structural property.

They do not claim that the calculated expansion modes reect the details of the evolutionary development during natural selection of metabolic pathways. This particularly concerns the results obtained using expansion condition C_1 in which many intermediate generations depend on the abundance of external metabolites. This problem is less severe when using expansion condition C_2 since networks get expanded only by such reactions which further convert metabolites made available by already existing reactions of the network.

9

Modularity in Metabolic Networks

The identification and characterisation of system-level features of biological organisation is a key issue of postgenomic biology. The concept of modularity assumes that cellular functionality can be seamlessly partitioned into a collection of modules. Each module is a discrete entity of several elementary components and performs an identifiable task, separable from the functions of other modules.

Spatially and chemically isolated molecular machines or protein complexes (such as ribosomes and flagella) are prominent examples of such functional units, but more extended modules, such as those achieving their isolation through the initial binding of a signaling molecule, are also apparent. Simultaneously, it is recognised that the thousands of components of a living cell are dynamically interconnected, so that the cell's functional properties are ultimately encoded into a complex intracellular web of molecular interactions. This is perhaps most evident with cellular metabolism, a fully connected biochemical network in which hundreds of metabolic substrates are densely integrated through biochemical reactions.

Within this network, however, modular organisation (i.e., clear boundaries between subnetworks) is not immediately apparent. Indeed, recent studies have demonstrated that the probability that a substrate can react with k other substrates [the degree distribution $P(k)$ of a metabolic network] decays as a power law $P(k) \sim \kappa^{\gamma}$ with $\gamma \cong 2.2$ in all organisms, suggesting that metabolic networks have a scale-free topology. A distinguishing feature of such scale-free networks is the existence of a few highly connected nodes (e.g., pyruvate or coenzyme A), which participate in a very large number of metabolic reactions.

With a large number of links, these hubs integrate all substrates into a single, integrated web in which the existence of fully separated modules is prohibited by definition (Fig. 1A). Yet, the dilemma of a modular versus a highly integrated module-free metabolic network organisation remains. A number of approaches for analysing the functional capabilities of metabolic networks indicate the existence of separable functional elements. Also, from a purely topologic perspective, the metabolic network of Escherichia coli is known to possess a high clusteringcoefficient, a property that is suggestive of a modular organisation. In itself, this implies that the metabolism of E. coli has a modular topology, potentially comprising several deely interconnected functional modules of varying sizes that are connected by few intermodule links (Fig. 1B). However, such clear-cut modularity imposes severe restrictions on the degree distribution, implying that most nodes have approximately the same number of links, which contrasts with the metabolic network's scale-free nature.

To determine whether such a dichotomy is indeed a generic property of all metabolic networks, first calculated the average clustering coefficient for 43 different organisms as a function of the number of distinct substrates N present in their metabolism. The clustering coefficient, defined as $C_i = 2n/k_i(k_i - 1)$, where n denotes the number of direct links connecting the ki nearest neighbors of node i, is equal to 1 for a node at the center of a fully interlinked cluster, and it is 0 for a metabolite that is part of a loosely connected group (Fig. 2A).

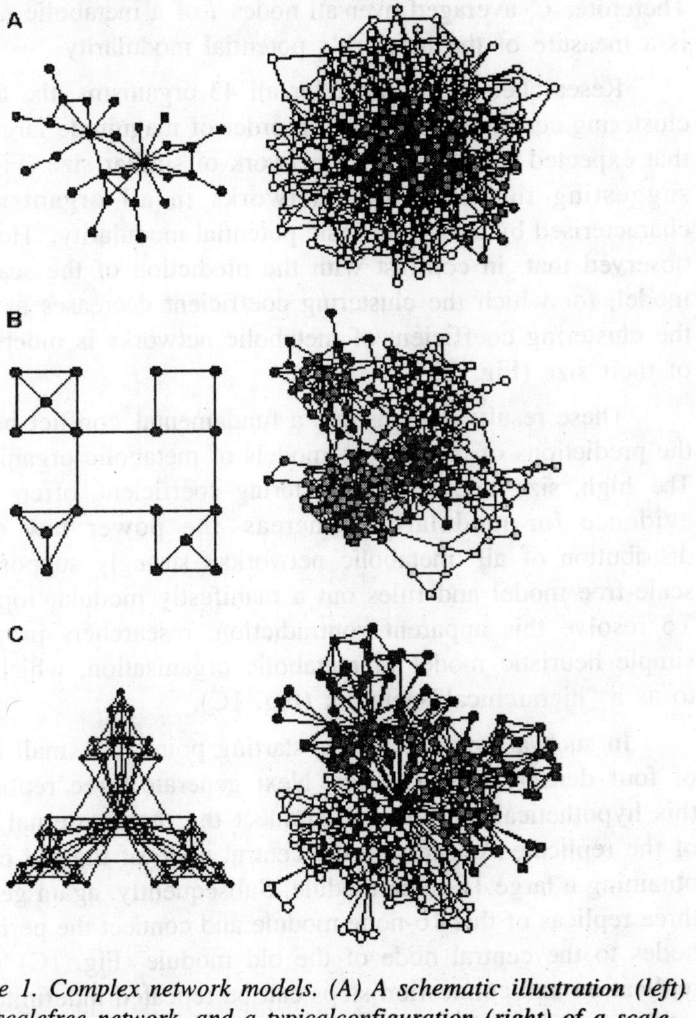

Figure 1. Complex network models. (A) A schematic illustration (left) of a scalefree network, and a typicalconfiguration (right) of a scale-free network (B) Schematic illustration (left) of a manifestly modular network made of four highly interlinked modules connected to each other by a few links. A standard clustering algorithm uncovers the network's inherent modularity (right) by partitioning a modular network of N = 256 nodes into the four isolated structures built into the system. (C) The hierarchicalnetwork (left) has a scale-free topology with embedded modularity. Standard clustering algorithms (right) are less successful in uncovering the network's underlying modularity.

Therefore, C_i averaged over all nodes i of a metabolic network is a measure of the network's potential modularity.

Researchers found that, for all 43 organisms, the average clustering coefficient is about an order of magnitude larger than that expected for a scale-free network of similar size (Fig. 2B), suggesting that metabolic networks in all organisms are characterised by a high intrinsic potential modularity. Here also observed that, in contrast with the prediction of the scale-free model, for which the clustering coefficient decreases as $N^{-0.75}$, the clustering coefficient of metabolic networks is independent of their size (Fig. 2B).

These results demonstrate a fundamental conflict between the predictions of the current models of metabolic organisation. The high, size-independent clustering coefficient offers strong evidence for modularity, whereas the power law degree distribution of all metabolic networks strongly supports the scale-free model and rules out a manifestly modular topology. To resolve this apparent contradiction, researchers propose a simple heuristic model of metabolic organisation, which refer to as a "hierarchical" network (Fig. 1C).

In such a model network, starting point is a small cluster of four densely linked nodes. Next generate three replicas of this hypothetical module and connect the three external nodes of the replicated clusters to the central node of the old cluster, obtaining a large 16-node module. Subsequently, again generate three replicas of this 16-node module and connect the peripheral nodes to the central node of the old module (Fig. 1C). These replication and connection steps can be repeated indefinitely, in each step quadrupling the number of nodes in the system. The architecture of such a network integrates a scale-free topology with an inherent modular structure. It has a power law degree distribution with degree exponent $\gamma = 1 + (\ln 4)/(\ln 3) = 2.26$, in agreement with $\gamma = 2.2$ observed in metabolic networks. Its clustering coefficient $C \cong 0.6$ is also comparable with that observed for metabolic networks.

Most important, the clustering coefficient of the model is independent of the size of the network, in agreement with the results of Fig. 2B. A unique feature of the proposed network model, not shared by either the scalefree scalefree (Fig. 1A) or modular (Fig. 1B) models, is its hierarchical architecture. This hierarchy, which is evident from a visual inspection, is intrinsic to the assembly by repeated quadrupling of the system. The hierarchy can be characterised quantitatively by using the recent observation that, in deterministic scale-free networks, the clustering coefficient of a node with k links follows the scaling law $C(k) \sim k^{-1}$. This scaling law quantifies the coexistence of a hierarchy of nodes with different degrees of modularity, as measured by the clustering coefficient, and is directly relevant to the model (Fig. 1C).

Indeed, the nodes at the center of the numerous 4-node modules have a clustering coefficient C = ¾, those at the center of a 16-node module have $k = 13$ and $C = 2/13$, and those at the center of the 64-node modules have $k = 40$ and $C = 2/40$, indicating that the higher a node's connectivity, the smaller its clustering coefficient, asymptotically following the $1/k$ law. To investigate whether such hierarchical organisation is present in cellular metabolism, and measured the $C(k)$ function for the metabolic networks of all 43 organisms.

Figure 2 shows the hierarchicalmodularity in metabolic networks. Figure 2(A) is the clustering coefficient offers a measure of the degree of inter connectivity in the neighbourhood of a node. For example, a node whose neighbours are all connected to each other has C = 1 (left), whereas a node with no links between its neighbours has C = 0 (right). Figure 2(B) is the average clustering coef.cient C(N) for 43 organisms is shown as a function of the number of substrates N present in each of them. The dashed line indicates the dependence of the clustering coefficient on the network size for a module-free scale-free network, and the diamonds denote C for a scale-free network with the same parameters (N and number of links) as observed in the 43 organisms. (C through E) The dependence of the

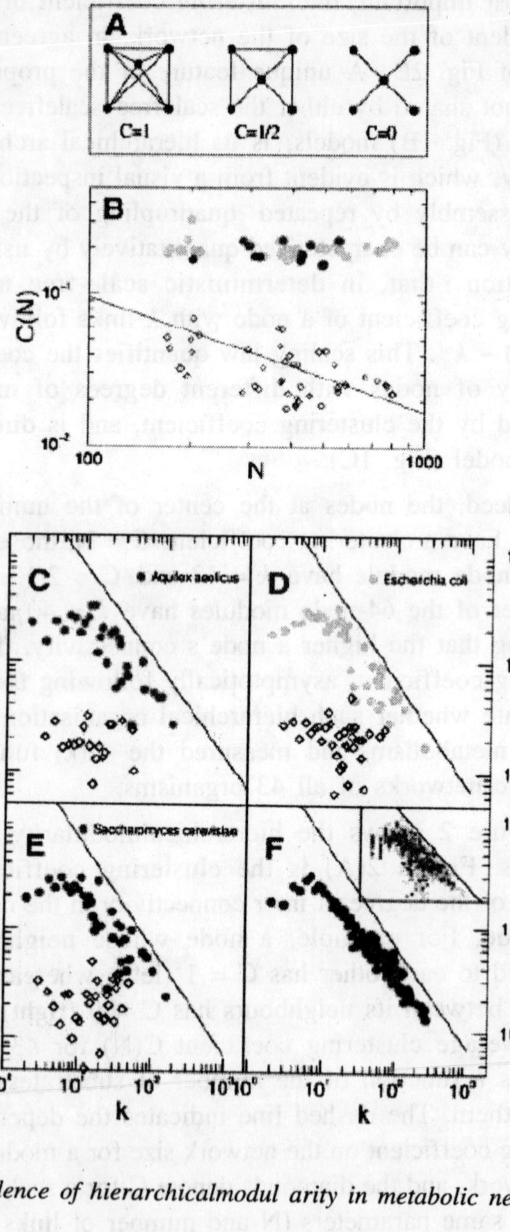

Figure 2. Evidence of hierarchicalmodul arity in metabolic networks.

clustering coef.cient on the node's degree in three organisms: Aquidex aeolicus (archaea) (C), Escherichia coli (bacterium) (D), and Saccharomyces cerevisiae (eukaryote) (E). Figure 2(F) is the C(k) curves averaged over all 43 organisms is shown, and the inset displays all 43 species together. In (C through F), the dashed lines correspond to $C(k) = k^{-1}$, and in (C through E), the diamonds represent the C(k) value expected for a scale-free network (Fig. 1A) of similar size, indicating the absence of scaling. The wide fluctuations are due to the small size of the network.

As shown in Fig. 2, C through F, for each organism, $C(k)$ is well approximated by $C(k) \sim k^{-1}$, in contrast to the k-independent $C(k)$ predicted by both the scale-free and modular networks. This provides direct evidence for an inherently hierarchical organisation. Such hierarchical modularity reconciles within a single framework all the observed properties of metabolic networks: their scale-free topology; high, system size-independent clustering coefficient; and the power law scaling of $C(k)$.

A key issue from a biological perspective is whether the identified hierarchical architecture reflects the true functional organisation of cellular metabolism. To uncover potential relations between topological modularity and the functional classification of different metabolites, researchers concentrated on the metabolic network of E. coli, whose metabolic reactions have been exhaustively studied, both biochemically and genetically. Using a previously established graph-theoretical representation, first subjected E. coli's metabolic organisation to a three-step reduction process, replacing nonbranching pathways with equivalent links, allowing us to decrease its complexity without altering the network topology.

The organisation of metabolic networks is likely to combine a capacity for rapid flux reorganisation with a dynamic integration with all other cellular function. The system-level structure of cellular metabolism is best approximated by a hierarchical network organisation with seamlessly embedded modularity.

There are many highly integrated small modules, which group into a few larger modules, which in turn can be integrated into even larger modules. This is supported by visual inspection of the derived hierarchical tree, which offers a natural breakdown of metabolism into several large modules, which are further partitioned into smaller, but more integrated submodules. The mathematical framework proposed here to uncover the presence or absence of such hierarchical modularity and to delineate the modules based on the network topology could apply to other cellular and complex networks as well.

As scale-free topology has been found at many different organisational levels, ranging from genetic to protein interaction and protein domain networks, it is possible that biological networks are always accompanied by a hierarchical modularity. Some nonbiological networks, ranging from the World Wide Web to the Internet, often combine a scale-free topology with a community structure (i.e., modularity); therefore, these networks are also potential candidates for hierarchical modularity.

For biological systems, hierarchical modularity is consistent with the notion that evolution may act at many organisational levels simultaneously: The accumulation of many local changes, which affect the small, highly integrated modules, could slowly impact the properties of the larger, less integrated modules. The emergence of the hierarchical topology through copying and reusing existing modules and motifs, a process reminiscent of the results of gene duplication, offers a special role to the modules that appeared first in the network. Although the model of Fig. 1C reproduces the large-scale features of the metabolism, understanding the evolutionary mechanism that explains the simultaneous emergence of the observed hierarchical and scale-free topology of the metabolism, as well as its generality to cellular organisation, is now a prime challenge.

10

CONNECTIVITY IN METABOLIC NETWORKS

Metabolic networks belong to the best-studied biochemical networks. It consist of metabolites and of biochemical reactions transforming these metabolites into each other. In contrast to other biochemical networks such as signal transduction networks and genetic networks, complete network topologies can be easily obtained from annotated genomes.

The connectivity of metabolic networks follows approximately a power law, i.e., the frequency, P(k), of metabolites participating in k reactions is given by $P(k) \sim k^{-\gamma}$, where γ is a constant coefficient. This distribution implies that although most metabolites are involved in only few reactions, some metabolites serve as "hubs" and are involved in many reactions.

These hubs are more frequent then would be expected, for example, in random networks. Interestingly, many hubs, such as ATP, NADH, glutamate, coenzyme A, and their derivates, serve as key compounds in the transfer of specific biochemical groups, such as phosphate groups, redox equivalents, amino groups, and acetyl groups.

Using computer simulations they here study whether hubs emerge in a widely accepted scenario for the evolution of metabolic networks. This scenario has originally been proposed by Kacser and Beeby and considers an early stage in the evolution of life. It is based on the assumption that cells exist with membranes that separate the interior of the cell from the environment, a metabolism that allows cells to transform compounds taken up from the environment into those that are required for the formation of biomass, and genes coding for enzymes that catalyze these biochemical reactions.

Because of the low coding capacity and because it is unlikely that a large number of specialised enzymes emerge de novo, it is plausible to assume that these ancestral cells had only a few enzymes with broad specificities. The broad specificity allowed catalysing all essential reactions at the cost of a low turnover for any single biochemical reaction. The ancestral cells are assumed to have been selected for growth rate and evolve by mutations affecting the kinetic properties of the enzymes and occasional gene duplications.

This lead to the emergence of networks with highly specialised enzymes as observed in modern organism.A number of alternative scenarios for the evolution of novel enzymes and metabolic pathways have been proposed. One scenario for the evolution of metabolism in the early history of life has been proposed by Horowitz and is referred to as "backwards evolution." This scenario starts in an environment containing a large number of metabolites.

Therefore only a low number of reactions are required to produce a useful product from one of these metabolites. As metabolites get depleted from the environment, longer pathways evolve in a reverse fashion, connecting the product to more distant metabolites present in the environment. A further scenario for the evolution of novel enzymes and metabolic pathways is referred to as "enzyme recruitment".

However, it does not apply to the initial emergence of metabolic networks early in the history of life, because it assumes that a core metabolic network with specialised enzymes is already present. Novel pathways emerge by recruiting enzymes from existing pathways. Analysis of distribution patterns of homologous enzymes in metabolic networks and of reactions catalysed by different enzymes of the same family support enzyme recruitment rather than backwards evolution.

This scenario represents a plausible mechanism for the evolution of metabolic networks in early organisms. It remains to be shown whether it indeed leads to the emergence of metabolic networks with connectivity distributions similar to those observed in nature. Based on the scenario of Kacser and Beeby, they therefore develop a model that allows simulating the evolution of metabolic networks and compare the resulting networks with natural networks.

Thomas Pfeiffer et. al conducted a study on metabolic networks to determine the evolution of connectivity. Details of their study are discussed below.

They simulate the evolution of metabolic networks based on the following assumptions:Because of the importance of group transfer in metabolic networks they assume that metabolic networks are group transfer networks (Figure 1). Metabolites consist of biochemical groups. Enzymes catalyse the transfer of groups between metabolites.

Additionally there are transporters that catalyse the uptake or excretion of metabolites. Some of the metabolites are involved in the formation of biomass. The concentrations of these metabolites determine the rate of biomass formation. Biomass formation results in growth. The growth rate is given by the rate and the costs of biomass formation.

The dynamics of the metabolites is described by a set of ordinary differential equations for the metabolite concentrations and is based on the kinetic properties of the reactions and transport processes. Additionally, growth leads to the dilution of metabolites at a rate equal to the growth rate. The system has therefore no conservation relations, and metabolites that are not involved in a reaction or transport process have a steady-state concentration of zero.

There are two ways in which a metabolite can have a non-zero steady state. Either it participates in at least two reactions-one of which produces the metabolite, while the other one consumes the metabolite-or a metabolite is produced by a single reaction and is diluted due to growth. Fitness is assumed to be proportional to the growth rate at steady state of the system. Thus fitness is a dynamic property of the entire network.

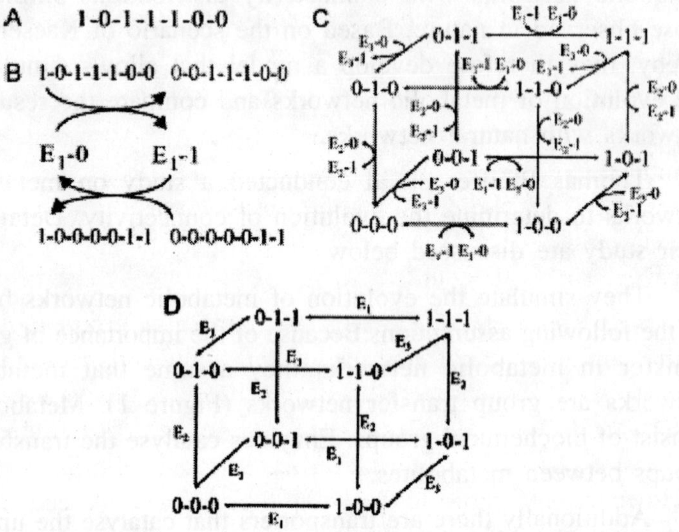

Figure 1. Structure of the Initial Network(A)

In figure 1(A) we can see metabolites are assumed to consist of N different biochemical groups. Metabolites can be denoted as binary strings, because each group can be either present or absent. In the example given, the metabolite carries groups 1, 3, 4, and 5 whereas all other groups are absent.

In figure 1(B), enzymes catalyse the transfer of specific biochemical groups between metabolites. For the group transfer they assume a ping-pong-like mechanism that consists of two half-reactions. In the first half-reaction, the group is transferred from a donor to the enzyme. Thereby the donor is transformed into its corresponding acceptor. In the second half-reaction, the enzyme transfers the group to another acceptor, thereby transforming it into its corresponding donor.

They assume that a half-reaction follows linear reversible kinetics. This results in Michaelis-Menten-like kinetics for the transfer of a group from one metabolite to another. For the initial network they assume that enzymes are specific with regard to the group but unspecific with regard to donors and acceptors, i.e., they transfer a specific biochemical group from any donor to any acceptor.

In figure 1(C) the initial network consists of 2N metabolites. A single half-reaction transforms the metabolites into one of the N neighbours, such that the initial network resembles an N-dimensional hypercube. Note, however, that for the transfer of a group, two edges (i.e., half-reactions) in the cube have to be coupled.

In figure 1(D), to study the impact of group transfer on the connectivity distribution of the simulated networks, they performed simulations with monomolecular reaction networks, in which groups are added and removed, but not transferred, by the enzymes. The initial network has a similar structure. However, in contrast to the group transfer network, each reaction can be performed without coupling to another reaction.

Initially all enzymes and all transporters are unspecific and allow small mutations in their kinetic properties. These mutations increase the rate constant of a single reaction while decreasing the rate constants of all other reactions of an enzyme or transporter. Alternatively they decrease the rate constant of a single reaction while increasing the rate constants of the other reactions.

In addition to mutations affecting kinetic properties, they assume that there are gene duplications (or deletions) that increase (decrease) the dosage of enzymes and transporters. Changing the dosage of an enzyme has two opposing effects on fitness: An increased dosage increases the fluxes in the network, but it is also associated with increased metabolic costs of biomass formation. Duplications also enable enzymes to diverge and specialise on different reactions.

If a multifunctional enzyme has only a small impact on biomass formation, duplication can be detrimental provided that the metabolic costs of the additional copy are larger than the benefits in terms of increased growth rate. Consequently, enzymes and transporters do not necessarily specialise completely over the course of evolution.

On the other hand, duplications can be beneficial if enzymes or transporters are already completely specialised. Therefore multiple identical copies of completely specialised enzymes and transporters can evolve provided they have a large impact on biomass formation. In their simulations, they assume that there are seven different biochemical groups that characterise each metabolite.

Because a metabolite can either carry one or no copies of each group, they have an initial network of 128 metabolites. The simulations with seven unspecific enzymes that catalyse the transfer of one of the biochemical groups from any donor to any acceptor, and with a single unspecific transporter that catalyses the transport of all metabolites. Sixteen randomly chosen metabolites are involved in biomass formation.

For the production of biomass, it is necessary to have at least one donor and acceptor of each group as external metabolites. They therefore assume that a minimal set of two metabolites, namely X0 and X127, is present in the environment. For reasons of computational tractability, they let the network evolve by fixation of the most beneficial mutations rather than random mutations. Duplications and deletions, because they assume that they are rare compared to mutations that alter the kinetic properties, are only considered if no further beneficial kinetic mutations are possible.

RESULTS

A sample simulation is shown in Figure 2. The simulation

indicates that the scenario proposed by Kacser and Beeby indeed leads to the evolution of metabolic networks with specialised enzymes (Figure 2A). The emerging network consists of 23 different reactions and seven transport processes that are catalysed by fully specialised enzymes and transporters (Figure 2C). The enzymes and transporters are present in multiple identical copies.

The network contains only a fraction of the metabolites present in the initial network. The remaining metabolites become unconnected as the reactions specialise, i.e., they do not serve as substrate or product of reactions present in the emerging network. Figure 2B shows the connectivity distribution of the metabolites in the emerging network.

The connectivity includes participation in enzymatic reactions, transport processes, and biomass formation. Most metabolites have a connectivity of two. In this simulation there are no metabolites that participate in only one reaction or transport process. Some of the metabolites, such as X0, X16, X18, X126, and X127, are involved in a comparably large number of reactions.

These metabolites "monopolize" the transfer of specific biochemical groups. One of the groups, for example, is always transferred from metabolite X127 to a number of different acceptors. Hence, it seems to be beneficial that at least some of the metabolites specialise in the transfer of specific groups by acting as the only donor or acceptor.

Figurte 2(A), the initial network consists of 128 metabolites, seven unspecific enzymes (each of which transfers one of the seven biochemical groups that metabolites carry), and a single unspecific transporter. Within the course of evolution, the enzymes and transporter duplicate and increase in specificity (i.e., the number of half-reactions per enzyme and of metabolites per transporter decreases).

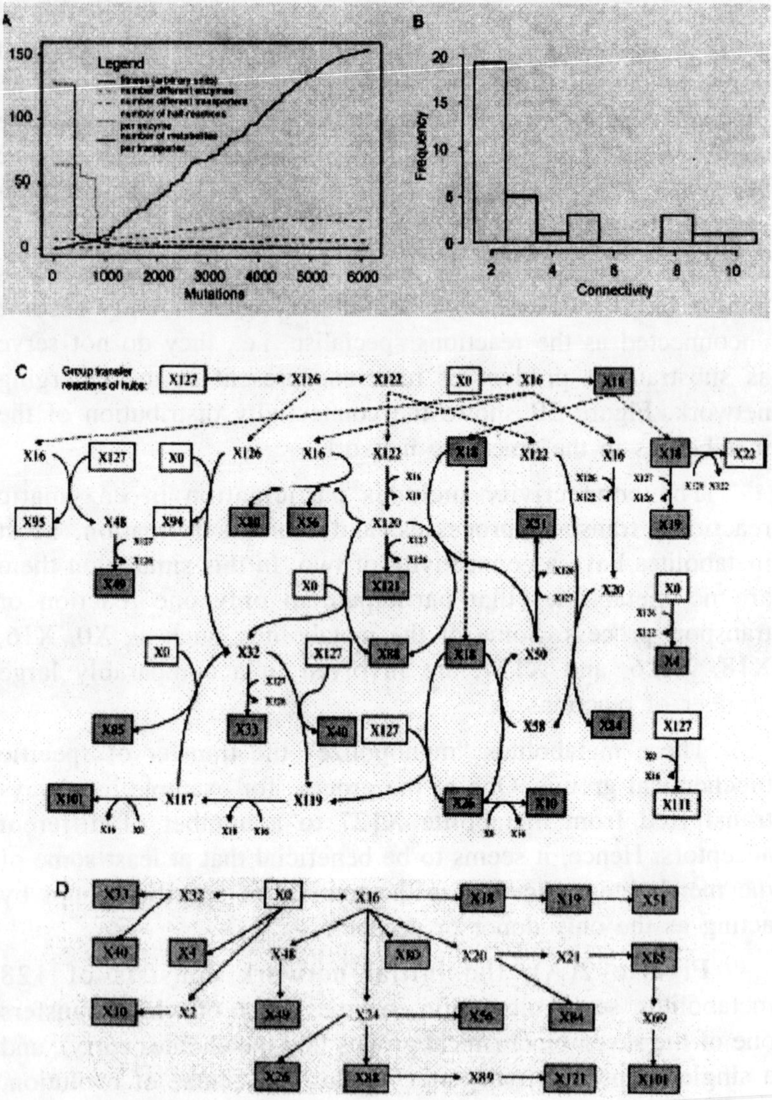

Figure 2. Sample Simulation for the Evolution of Metabolic Networks

The emerging network consists of 23 enzymatic reactions and seven transport processes. In the sample simulation, all enzymes and transporters in the emerging network are highly specific, i.e., the enzymes catalyse only two half-reactions, and the transporters transport single metabolites. The emerging network contains only 33 metabolites. The remaining metabolites are not involved in the emerging network.

Figurte 2(B) shows the connectivity distribution of the emerging group transfer network. Most metabolites are involved in only two reactions. However, a few metabolites are highly connected.

Figurte 2(C) shows the pathway scheme of the emerging group transfer network. The metabolites X0 and X127 are taken up from the environment, whereas metabolites X4, X22, X94, X95, and X111 are excreted into the environment (white boxes). The network eventually transforms metabolites X0 and X127 into those metabolites that are involved in biomass formation (grey boxes). Interestingly, metabolite X4 is excreted although it is involved in biomass formation.

Note that some half-reactions evolve, such as the one from X127 to X126, and monopolise the transfer of a specific group (in this case the first group in the binary string). These metabolites are involved in many reactions and therefore have high connectivity. The group transfer reactions of these hubs are summarised in the first line of the pathway scheme. The emerging group transfer network is much more complex than the corresponding monomolecular reaction network (Figure 2D) and even includes a cycle (X32→X119→X117→X32), with the net reaction of X0 + X16 + X127→X18 + X40 + X85).

Figure2(D) shows pathway scheme of an emerging monomolecular reaction network. Although the network produces the same set of metabolites involved in biomass formation as the group transfer network shown in Figure 2C, it has a much simpler, tree-like structure. Additionally, the network has no obvious hubs.

To obtain more precise data for the connectivity distribution, they analysed 40 replicate simulations. The results of these simulations are similar to the sample simulation shown in Figure 2. Most of the enzymes and transporters of the emerging networks specialise completely, i.e., catalyse two half-reactions or the transport of a single metabolite. However, a small fraction of the enzymes in the emerging networks catalyse more than two half-reactions, i.e., are not fully specialised. These enzymes have only a weak impact on the rate of biomass formation and therefore cannot duplicate and specialise fully.

Most of the fully specialised enzymes and transporters are present in multiple identical copies. As in the sample simulation, the emerging networks contain only a small fraction of the metabolites present in the initial networks. Most of these metabolites are involved at least in two reactions or transport processes.

Occasionally they observe metabolites that participate in only a single reaction. Figure 3A shows the connectivity distribution over all emerging networks. Metabolites that are participating in fewer than two reactions are not included in the graph. The average maximum likelihood estimate for the power-law coefficients of the individual networks is $\gamma = 2.63$ (± 0.17).

However, the overall connectivity distribution deviates strongly from a power law (Figure 3B). In particular, the overall distribution has more hubs than expected in the connectivity range between eight and 12, but has fewer hubs than expected with a connectivity of more than 12. This disagreement between the emerging networks and natural networks may arise because the networks emerging in their simulations are much smaller than the networks analysed in the literature.

They therefore tested whether the simulated networks are consistent with subnetworks of similar size derived from real metabolic networks. These networks were obtained from the KEGG database. The KEGG database displays several functional metabolic subnetworks such as glycolysis, the tricarboxylic acid cycle, or amino acid synthesis for a wide range of organisms.

They analysed 22 subnetworks from Escherichia coli, each containing between 24 and 40 metabolites. The overall connectivity of these networks is shown in Figure 3C. Note, that in order to compare the connectivity distributions of the simulated and the natural networks, they treat the natural networks as independent networks, i.e., a metabolite that is present in two different networks is treated as two different metabolites. As in the simulated networks, they consider only metabolites that are involved in more than one reaction.

A comparison of the connectivity distribution of the natural and the emerging networks indicates that both connectivity distributions are not significantly different. This result indicates that although the connectivity of the networks emerging in their simulation is not consistent with that of large metabolic networks described in the literature, it is similar to the connectivity observed in natural metabolic subnetworks of the same size.

Note, however, that these subnetworks typically do not produce all compounds required for biomass formation and therefore have been selected for functions different from the networks emerging in their simulations.

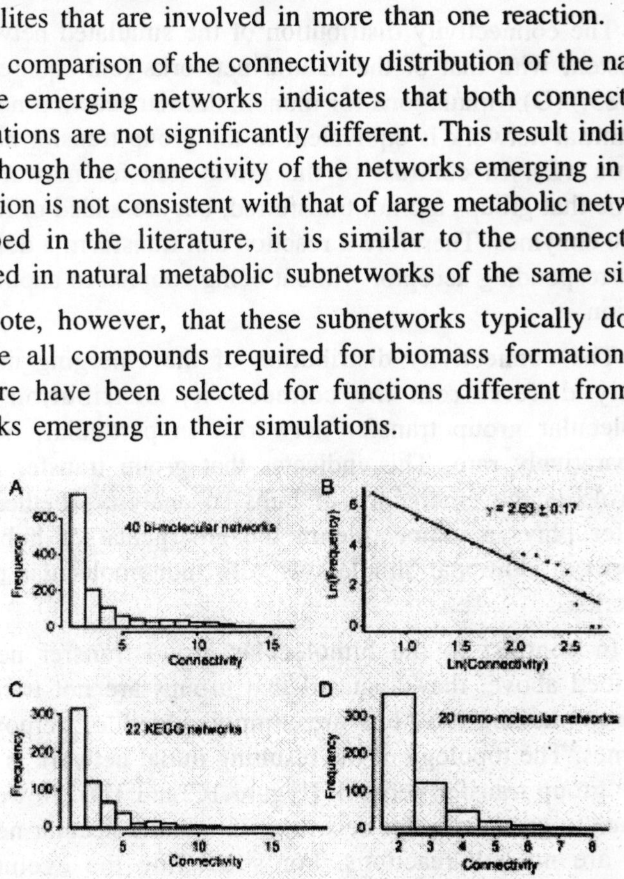

Figure 3. Connectivity Distribution of Simulated and Natural Networks

In the above figure (A) is the connectivity distribution of 40 different simulated networks. As in the sample simulation, most metabolites are only involved in two or three reactions. On the other hand, the distribution shows that there are a number of hubs that are involved in up to 16 reactions.(B) The double-logarithmic plot of the distribution reveals a considerable deviation from a power-law distribution. (C) Connectivity distribution in natural metabolic networks of similar size. The networks are generated from the KEGG database for E. coli.

The connectivity distribution of the simulated networks is consistent with that of the natural networks (chi square test: $p \approx 0.085$;).(D) Simulations for monomolecular reaction networks. The initial network is equivalent to the group transfer networks (Figure 1C). In contrast to the group transfer network, they assume that groups are not transferred, but are added or removed by the enzymes. Therefore a reaction can transform a donor into its corresponding acceptor without being coupled to another half-reaction.

The connectivity distribution of the emerging networks clearly differs from the connectivity distribution in the bimolecular group transfer networks. In particular, hubs are comparatively rare. This indicates that group transfer plays a key role in the emergence of hubs.To analyse whether group transfer plays a major role in the emergence of hubs, they performed additional simulations with monomolecular reaction networks.

In contrast to the bimolecular group transfer networks described above, they assumed that groups are not transferred between metabolites, but are simply added or removed by enzymes. The topology of the resulting initial network is similar to the group transfer network (Figure 1C and 1D). However, in contrast to group transfer networks, in monomolecular networks there are no half-reactions. For simulating the evolution of monomolecular networks, they used assumptions analogous to those used for the simulations of the bimolecular group transfer networks. This network produces the same set of metabolites involved in biomass formation as the network shown in Figure 2C. However, the resulting monomolecular reaction network differs remarkably from the bimolecular reaction network. In contrast to the group transfer reactions, monomolecular reactions have only a single substrate and product. Therefore monomolecular networks can be displayed as graphs that typically have a tree-like structure (Figure 2D).

The connectivity distributions of the simulated monomolecular networks clearly differs from the simulated group transfer networks (Figure 3D). In particular, monomolecular networks have much fewer hubs. Their simulation results indicate that group transfer plays a key role in the emergence of hubs. Within the evolution of group transfer networks, hubs emerge because some metabolites become a preferential donor or acceptor of a specific group, i.e., they have a particularly high group transfer potential.

Therefore it is of advantage that an enzyme that transfers this specific group increases its specificity towards this donor (or acceptor). Once a preferential donor (or acceptor) emerges and the corresponding enzyme starts to specialise towards this metabolite, the remaining enzymes in the network are selected for further increasing the group transfer potential of this specific metabolite.

Thereby a single metabolite may "monopolize" this group transfer reaction. It serves as the only donor (or acceptor) of a specific group towards a number of different acceptors (donors) in the network. As a consequence it participates in a number of different reactions. In the absence of group transfer reactions, i.e., in monomolecular reaction networks, this mechanism does not work. Therefore, hubs do not emerge in monomolecular reaction networks.

Their simulations demonstrate that metabolic networks may have emerged according to the scenario of Kacser and Beeby. Starting from metabolic networks with multifunctional enzymes, networks emerge that consist of highly specialised enzymes. The key evolutionary mechanism in this scenario is the duplication of enzymes followed by the specialisation of the copies for different biochemical reactions. The implication of this mechanism is that during the early evolution of metabolic networks, biochemical reactions and intermediate metabolites have been lost.

On the basis of their model, they expect that multifunctional enzymes can be maintained if they have only little impact in biomass formation and are therefore present as single copies. Alternatively, multifunctional enzymes can be maintained if there is no trade-off between the different functions. Recent results on the evolution of multifunctional enzymes indicate that this may at least be sometimes the case.

Although the overall connectivity of the networks emerging in their simulations differs clearly from a power-law distribution, it is consistent with the connectivity distribution derived from natural subnetworks of similar size. Their simulations indicate that the scenario of Kacser and Beeby leads to the emergence of hub metabolites. Additionally, their results predict that group transfer is essential for the emergence of hubs in these networks.

This prediction is in line with the observation that most hubs in natural networks such as ATP, NADH, glutamate, and coenzyme A are key compounds in the transfer of biochemical groups. The high connectivity of these metabolites results mainly from transferase reactions rather than other reaction types.

In their simulations, hubs emerge as a consequence of selection for growth rate. In addition, along with a number of further publications given below, their simulations therefore support the view that hubs are not necessarily a result of selection for robustness as has been suggested previously. First, it has been shown that the evolution of robustness against complete knock-outs of genes, as for example by duplicates, requires very specific conditions in terms of mutation rates, gene functions, and interactions.

Second, a recent study on robustness and enzyme indispensability in yeast metabolism indicates that the apparent dispensability of many enzymes is not due to network robustness, but to the fact that many enzymes are only required under specific environmental conditions. Third, robustness against environmental changes is also unlikely to explain the connectivity distributions observed in natural networks.

This is because power-law connectivity distributions have been observed in a wide range of organisms living in very different environments, including, for example, intercellular parasites that live in very stable environments. Finally, no evolutionary scenarios have been presented that demonstrate that selection for increased robustness leads to the emergence of metabolic networks with power-law connectivity.

Nevertheless, it has been shown that robustness of metabolic networks against partially deleterious mutations may emerge as a "side product" of selection for growth rate. It can therefore be expected that the networks emerging in their simulations are robust against slightly deleterious mutations.

Metabolites

Metabolites are assumed to consist of seven different biochemical groups. Each biochemical group is either present once or is absent. Thus there are 2^7 different metabolites that are characterised by a binary string of length 7, where "1" at position x denotes the presence of group x, whereas "0" denotes the absence of the group.

Metabolites are associated with a random free energy that is taken from a uniform distribution between zero and one, and is required to specify thermodynamic properties of the biochemical reactions. For the production of biomass, it is necessary to have at least one donor and acceptor of each group as external metabolites. Thus a minimal set of two metabolites, namely X0 (characterised by the string 0-0-0-0-0-0-0) and X127 (characterised by the string 1-1-1-1-1-1-1), is assumed to be present in the environment. Sixteen randomly chosen metabolites are involved in biomass formation.

Enzymes

Enzymes catalyse the transfer of a specific biochemical group. They assume that groups are transferred by a "ping-pong mechanism" (Figure 1): A donor of a group transfers the group to the appropriate enzyme and is thereby transformed into its corresponding acceptor. The enzyme then transfers the group to an acceptor, thereby transforming it into its corresponding donor. Thus an enzyme can be in two possible states, $E_i^{(1)}$ and $E_i^{(0)}$, with $E_i^{(1)} + E_i^{(0)} = E_i$, where E_i is the total dosage of enzyme i, $E_i^{(0)}$ is the concentration of the enzyme without its group being bound to it, and $E_i^{(1)}$ is the concentration of enzyme with its group being bound to it.

The free energy difference between both states of an enzyme is assumed to be a random value taken from a uniform distribution ranging from zero to one. They further assume that in principle all metabolites that contain a specific biochemical group can serve as a donor for group transfer, whereas all metabolites that do not contain the group can in principle serve as an acceptor.

Linear kinetics for the transfer of a group to an enzyme, given by $\nu = k_{ij}(E_i^{(0)}X_{(ij)}^{(1)} - E_i^{(1)}X_{(ij)}^{(0)}/q_{ij})$, where k_{ij} is the rate constant of the reaction j of an enzyme i, $X_{(ij)}^{(1)}$ is the concentration of the donor of the reaction, $X_{(ij)}^{(0)}$ is the concentration of the corresponding acceptor, and q_{ij} is the equilibrium constant of the reaction resulting from the free energies of the reactants.

For the transfer of a group from a donor to an acceptor, two half-reactions need to be coupled. This results in Michaelis-Menten-like kinetics and implies that functional enzymes need to maintain nonzero rate constants for at least two reactions. They assume that initially the enzymes are unspecific and transform groups from each donor to each acceptor with the same rate constant. The initial dosage of enzymes is $E_i = 1$.

Transporters

Transporters transport metabolites passively across the cell membrane. The rate of transport is given by $v = T_i t_{ij}(X_j - X_j^{ext})$, where T_i is the dosage of the transporter i,tij is the rate constant of the transport of metabolite j,X_j is the metabolite concentration in the cell, and X_j^{ext} is the metabolite concentration in the environment.

They assume that in the initial network there is a single transporter that transports all metabolites with the same rate constant. The initial dosage of the transporter is $T_0 = 1$.

Biomass Formation, Growth, and Network Fitness

Biomass is formed by the condensation of specific metabolites. The rate of biomass formation follows linear kinetics given by the product $v_{BM} = K_{BM} \Pi_i X_i$ over all metabolites Xi that are involved in biomass formation. The rate constant is set to kBM = 1 in all simulations. They assume that the formation of biomass leads to growth. The growth rate is given by W = 1/VdV/dt = $v_{BM}/(C_0 + C_E E + C_T T)$, where C_0 is the amount of biomass that is required for structural compounds (i.e., those compounds that are not directly involved in cellular metabolism), CE is the amount of biomass per enzyme, C_T is the amount of biomass per transporter, E is the total dosage of enzymes, and T is the total dosage of transporters.

The parameters C_0, C_E, and C_T are set to 50, 1, and 1, respectively. Note that due to cell growth, metabolites are constantly diluted at a rate equal to the growth rate. The fitness of a network is assumed to be proportional to the steady-state growth rate.

Tradeoff between Specificity and Catalytic Activity

In line with the scenario of Kascer and Beeby, they assume that reactions can either catalyse a large number of reactions but with low activity for each enzyme, or a lower number or reactions with improved catalytic activities. Specifically, they assume that the sum $\Sigma_j k_{ij}^{1/\alpha}$ and $\Sigma_j t_{ij}^{1/\alpha}$ over all rate constants k_{ij} or t_{ij} of an enzyme or transporter, i, respectively, is constant. For values of $\alpha > 1$ increasing the rate constant for a single reaction has an over-proportional effect on all other rate constants. In their simulations they use $\Sigma_j k_{ij}^{1/\alpha} = 1$, $\Sigma_j t_{ij}^{1/\alpha} = 1$, and $\alpha = 2$.

This implies, for example, that a transporter catalysing the transport of a single metabolite has a four times higher rate constant for this reaction than a transporter that is specialised on the transport of two metabolites.

Mutations and the Course of Evolution

There are two types of mutations: (1) mutations that change the kinetic properties of an enzyme and (2) mutations that change the number of copies of an enzyme, i.e., gene deletions and duplications. For the first type of mutation they assume that the value of $k_{ij}^{1/\alpha}$ or $t_{ij}^{1/\alpha}$, respectively, for a single reaction is either increased or decreased by a small value of $m = 0.01$, while the rate constants of the other reactions are decreased, or increased appropriately.

Gene deletions decrease the dosage of an enzyme. Duplications increase the dosage of an enzyme. In their simulations they calculate the effect of all possible mutations of the current network on the steady-state growth rate to obtain the mutant with maximal increase in fitness. This mutation is then assumed to become fixed and the resulting network is used to search for the next mutations. The steady states are calculated using the NAG library function c05nbc for root finding of systems of nonlinear equations.

The solver may fail (or may obtain steady states with negative concentrations) if the initial approximation of the steady state is far away from the actual steady state of the system. In this case they integrated the system towards its steady state (using the NAG library function *d02ejc*) until the root finder succeeds. To calculate the effect of a mutation on the steady state, they solve the mutated system using the metabolite concentrations of the non-mutated system as initial approximation, because for small mutations, the steady state of the mutated systems can be expected to be close to the non-mutated system.

Note, however, that in a system of nonlinear equations there may be multiple steady states, unstable steady states, and saddle points. They therefore test whether the steady state of the most beneficial mutation obtained by the above procedure (i.e., the mutation that is assumed to become fixed) is reached by integration from initial metabolite concentrations close to zero.

This test failed in ten out of 50 simulations. These simulations are excluded from further analysis. Gene duplications and deletions are assumed to be rare compared to mutations affecting the catalytic properties of enzymes and transporters. They therefore consider these types of mutation only if none of the mutations affecting kinetic properties are beneficial. A simulation ends if there are no beneficial kinetic mutations nor gene duplications or deletions.

Analysis of E. coli Metabolic Subnetworks

The connectivity in metabolic subnetworks of E. coli downloaded from the KEGG database on the basis of the annotation in the reaction database. Metabolites participating in fewer than two reactions are excluded. For comparison with the networks emerging in their simulation, they included 22 subnetworks of similar size, i.e., with between 24 and 40 metabolites. To compare the connectivity distributions, they use Pearson's chi square test. Data from the tail of the distributions ($k > 11$) are put into a single class.

Simulation of Monomolecular Networks

For simulating the evolution of monomolecular networks, they assume that enzymes add or remove biochemical groups, thereby transforming a donor of a group into its corresponding acceptor and vice versa. In contrast to bimolecular reactions, it is not required to couple two half-reactions. Completely specialised enzymes catalyse a single reaction rather that two half-reactions.

The rate of a reaction is given by $v = E_i k_{ij}(X_{(ij)}^{(1)} - X_{(ij)}^{(0)}/q_{ij})$, where k_{ij} is the rate constant of the reaction j of an enzyme i, E_i is the dosage of the enzyme, $X_{(ij)}^{(1)}$ is the concentration of the donor, $X_{(ij)}^{(0)}$ is the concentration of the corresponding acceptor, and q_{ij} is the equilibrium constant of the reaction resulting from the free energies of the reactants. All other assumptions are equal to those used in the simulations of the bimolecular group transfer networks.

Comparison of Metabolic Pathways

At a higher level enzymes group conceptually into metabolic pathways as part of a dynamic information processing system—substrates are processed by enzymes yielding other substrates.

30 Genomes

WIT-system

A convenient system to obtain (almost) complete genomic, and organisational information about microbial organisms is the WIT-system. WIT, provides researchers with DNA and protein sequence information of complete or partially sequenced genomes. This information is associated with organisational information, gene- and operon-organisation and especially information about metabolic networks. A major goal of WIT is to perform a so called metabolic reconstruction of microbial genomes.

Following 29 genomes of microbial origin and of one multicellular organism (C. elegans) are accessible via the WIT system. Currently 16 genomes are completely sequenced. The remaining 14 genomes are involved in ongoing sequencing projects and have to be accessed with care.

Table 1. Microbial Genomes + One Multicellular Organism

Code Organism	KD[a]	Size[kB][b]	#ORFs[2]	cs[c]
AG Archaeoglobus fulgidus	A	2178.40	2493	x[21]
TH Methanobacterium thermoaut.	A	1751.38	1866	x[32]
PH Pyrococcus horikoshii	A	1738.51	1825	x[20]
MJ Methanococcus jannaschii	A	1739.93	1797	x[2]
AA Aquifex aeolicus	B	1590.78	1744	x[5]
DR Deinococcus radiodurans	B	3261.20	3771	o[35]
EC Escherichia coli	B	4639.22	4289	x[1]
YP Yersinia pestis	B	4501.71	4296	o[40]
HI Haemophilus in.uenzae	B	1830.14	1846	x[8]
PA Pseudomonas aeruginosa	B	6286.26	6477	o[49]
NG Neisseria gonorrhoea	B	2063.17	1853	o[47]
NM Neisseria meningitidis	B	2157.54	1838	o[36]
RC Rhodobacter capsulatus SB1003	B	2024.62	2099	o[46]
HP Helicobacter pylori	B	1667.88	1547	x[45]
CJ Campylobacter jejuni	B	1644.03	2106	o[41]
CY Synechocystis sp.	B	3573.47	3226	x[19]
PG Porphyromonas gingivalis	B	2447.62	1832	o[39]
BB Borrelia burgdorferi	B	1503.21	1688	x[9]
TP Treponema pallidum	B	1138.82	946	x[11]
CA Clostridium acetobutylicum	B	4030.73	3967	o[12]
ML Mycobacterium leprae	B	2420.76	1568	o[42]
MT Mycobacterium tuberculosis	B	4411.53	3924	x[3]
MG Mycoplasma genitalium	B	580.07	532	x[10]
MP Mycoplasma pneumoniae	B	816.39	674	x[17]
PN Streptococcus pneumoniae	B	2104.82	1844	o[37]
ST Streptococcus pyogenes	B	1799.24	1599	o[48]
EF Enterococcus faecalis	B	3209.12	2967	o[38]
BS Bacillus subtilis	B	4214.81	4093	x[22]
SC Saccharomyces cerevisiae	E	12057.28	6125	x[13]

CE Caenorhabditis elegans	E	63729.49	10064	o[43]

[a]Kingdom: A... Archaea, B...Bacteria, E... Eucarya
[b]These numbers has to be taken with care by ongoing sequence projects
[c]Completely sequenced genomes are marked by an x

Aligning sequences to each other and measuring distances, e.g. BLOSUM and PAM similarity matrices, is a common approach to deduce a relationship between individual biopolymers.

Given a small metabolic pathway of a substrate which is processed by an enzyme. The global distance Δ between such pathways is deduced by the individual distances between substrates ΔS and enzymes ΔE. In Fig. 4 an example for a simple pathway with two functional roles (n = 2) is shown.

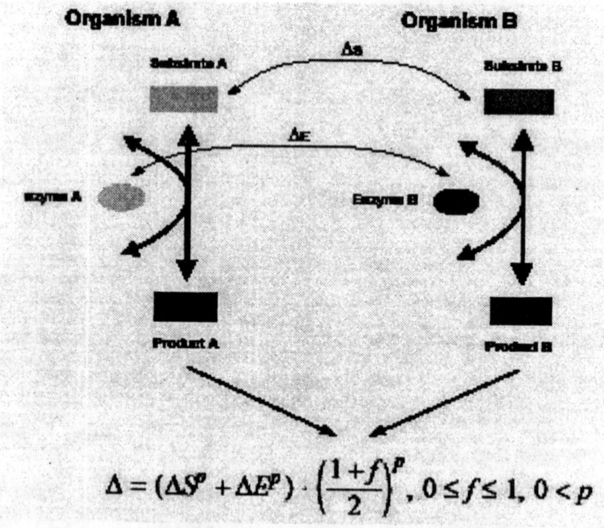

$$\Delta = (\Delta S^p + \Delta E^p) \cdot \left(\frac{1+f}{2}\right)^p, 0 \leq f \leq 1, 0 < p$$

Figure 4. Calculating a distance between two pathways: individual distances ΔE, ΔΣ between sequences of the same functional role are used to calculate a global distance Δ

A general definition for n functional roles per pathway is given in the following definition:

Let Γ, Γ′ be metabolic paths of identical topology involving n functional roles I_i, I'_i, i = 1... n. Let furthermore

$\Delta Xi = \delta(I_i, I'_i)$ be a distance calculated by an alignment δ. Then a distance Δ between Γ and Γ' is defined as:

$$\Delta = \Phi \cdot \sum_{i=1}^{n} \Delta X_i^p, \quad \Phi = \begin{cases} 1 & \textit{for orthologs} \\ \left(\frac{1+f(n-1)}{n}\right)^p & \textit{for paralogs} \end{cases}, \tag{1}$$

whereas $0 \leq f \leq 1$ and $p > 0$.

Paralog genes are discriminated against orthologs due to their (proposed) functional role in the genome which differs from functions of orthologs. Orthologs are genes in different species that evolved from a common ancestral gene by specification; by contrast, paralogs are genes related by duplication within a genome. Normally, orthologs retain the same function in the course of evolution, whereas paralogs evolve new functions, even if related to the original.

To reflect these differences for calculating distances between pathways parameter p is introduced in equ. 1. If $f = 1$ then orthologs and paralogs are treated the same. For $f = 0$ the total distance Δ between "paralog pathways" is actually the mean value of the individual distances ΔX_i. Exponent p in 1 is introduced to tune the discrimination between long and short distances. Large (positive) values of p emphasises large distances between individual functional roles.

Gaps

It is often meaningful to consider pathways with missing functional roles. This is especially the case when there are strong hints by experiments that a certain pathway is present in the corresponding organism. A simple but nevertheless useful approach is to introduce so called gap penalties for these missing functional roles:

Assume that in a pathway Γ a distinct functional role Ik is missing. Thus according to definition the distance ΔX_k is not defined which yields following definition of a gap penalty: A gap penalty is an arbitrarily assigned value Δ_{gap} to the otherwise undefined distance $\Delta_{gap} = \Delta_{\kappa} = \delta(.,I'_k)$.

This gap penalty must not be mistaken with the gap penalties used in sequence alignment. Additionally a threshold t as confidence level for the maximum number of accepted gaps can be defined. For the number of gaps exceeding t a pathway is no longer considered as valid.

Electron Transport

Electron transport pathways play a key role in the metabolism of a living cell. There are about 691 pathways known which are related to electron transfer. Table 2 previews a subset of 15 selected pathways out of the total 69. First let us define the following: A pathway Γ is defined to be present in the organism if the fraction of present functional roles for Γ exceeds a certain confidence level t. This, of course, is a formal definition and has always be confirmed by experiments. The confidence level t for pathways is 100%. Four members of this family are present in organisms of all three domains:

Dihydrolipoamide – NAD^+ Electron Transport (pathway 2)

NADH – FAD Electron transport (plasma membrane) (pathway 4)

NADH – NAD^+ Electron Transport (malate, aspartate) (pathway 10)

NADPH – Oxidised thioredoxin Electron transport (pathway 15)

The substrates utilised in these pathways also play functional roles in other metabolisms. E.g. lipoamide dehydrogenase component (E3) (EC 1.8.1.4) (pathway 2) catalyses reaction in carbohydrate metabolisms. NAD-dehydrogenase (EC 1.6.99.3) (pathway 4) is involved in many electron transport reactions. Aspartate aminotransferase (EC 2.6.1.1) and malate dehydrogenase (EC 1.1.1.37) (pathway 10) serve as reaction partners in aminoacid- and carbohydrate metabolisms. And last but not least thioredoxin and its reductase (EC 1.6.4.5) (pathway 15) is found in a variety of metabolism.

Thioredoxin reductase catalyses many electron transfer reactions (as two examples serve pathways 1, 15). Thioredoxin itself is involved in 29 pathways related to aminoacid-, electron transfer, protein, purine, pyrimidine and sulfur metabolisms as almost universal redox-reagens. This universality of thioredoxin and thioredoxin reductase is also reflected in its presence in almost all studied genomes (Table 1). This pathway is only missing in P. horikoshii, R. capsulatus, and C. elegans. The incomplete genomes of R. capsulatus and C. elegans lack the sequence of thioredoxin reductase. Wherease P. horikoshii does not possess thioredoxin.

On the contrary, many electron transfer pathways are only present in organisms of one or two domains. Specialised pathways involving mitochondria are of course found in Eucarya only (pathways 5, 9). A large number of pathways are assigned to Bacteria only (pathways 2, 7, 8, 12, 13, 14). For example the "pathological" mercury (II) reductase reaction (pathway 12) is utilised by four Bacteria only. There are no electron transport pathways which are unique for Archaea. This observation confirms the know fact that entire pathways are acquired or displaced in the case of the Archaea.

Actually, by investigating frequency distribution of functional roles between all three domains a tendency of horizontal gene transfer from Bacteria to Archaea can be detected. The reverse direction (from Archaea to Bacteria) is less probable. Studying pathway-"distribution" within a domain (here: Bacteria) shows the following. In general one can observe a coarse correlation between genome size (or number of ORFs) and number of present pathways. As exception to this rule serves B. burgdorferii which genome is almost as long as the genome of H. influenzae: In contrast to H. influenzae which uses many electron transfer pathways, highly parasitic B. burgdorferii posesses only NADPH – oxidised thioredoxin electron transport (pathway 15).

Without surprise, close related organisms use similar pathways: each pair of Neisseria NG, NM; Mycobacteria ML, MT; and Mycoplasmae MG, MP have the same set of pathways. Here the exception are Streptococcae, where S. pyrogens posesses a richer set of pathways than S. pneumoniae. From the family of Mycobacteria and Mycoplasmae, M. tuberculosis utilises unproportionally many pathways. A special state manifests the quadruple E. coli, Y. pestis, H. influenzae, and P. aeruginosa.

E. coli as one of the best studied microbial organism has the most complete set of pathways. Close related to this species the pathogens H. influenzae, Y. pestis and the metabolic very versatil P. aeruginosa possess a similar but less abundance of electron transfer pathways. In contrast to comparative genomics studies performed for E. coli and H. influenzae genomes there are no pathways present in pathogens which are missing in free living relatives.

Ferredoxin, an important Coenzyme

Ferredoxins, besides thioredoxin, flavodoxin and rubredoxin are important coenzymes in metabolic pathways. They serve as electron-acceptors and -donors in many anabolic, catabolic and electron transfer reactions. E.g. ferredoxin is redoxpartner in more than 50 known pathways as part of amino acid metabolism, aromatic hydrocarbons metabolism, carbohydrate metabolism, electron transfer, hydrogen metabolism, lipid metabolism, nitrogen metabolism, and sulfur metabolism. One quite important pathway in this class is the reversible Ferredoxin—NADPH Reductase pathway which can be found in Bacillaceae, Cyanobacteria and Enterobacteriaceae. In this redox-pathway ferredoxin is processed (either oxidised or reduced) by ferredoxin - NADPH reductase.

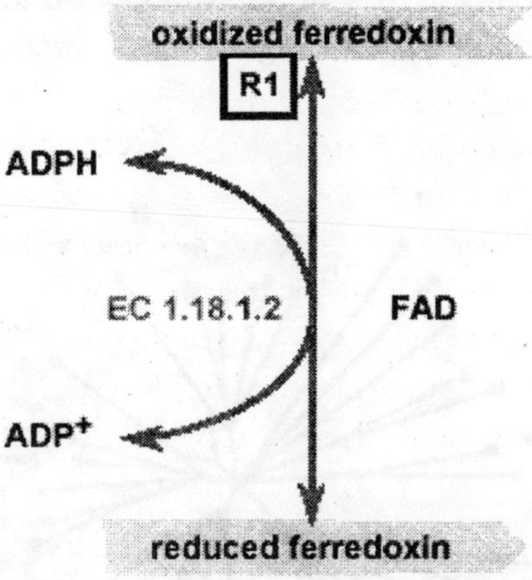

Figure 5. F erredoxin—NADPH reductase pathway: The pathway itself is shown with ferredoxin and ferredoxin—NADPH reductase (EC 1.18.1.2) as involved biopolymers.

Sequences for ferredoxin and its reductase are obtained from the WIT-system. A multiple sequence alignment for each set of sequences is performed by ClustalW v1.74 with the *BLOSUM62* similarity parameter. Alignment parameter are set to default values. The created Phylip distance matrices are then used for calcuiations of the pathway distance. Phylogenetic relationships are analysed by phylogenetic graph reconstruction programmes such as *SPLITSTREE2* or the *PHYLIP* software suite. Schematics of the pathway and a non-rooted phylogenetic graph drawn by *SPLITSTREE2* are shown in Fig. 5 and Fig. 6 respectively.

The nomenclature is according to theWITsystem. All pathways of the same organism are clustered in general. The cluster of Synechocystis sp. (*cyXX*) shows this property quite nicely. Similar clusters are observed for D. radiodurans (*drXX*) and M. tuberculosis (*mtXX*). On the contrary a more complex relationship between the Neisseria, P. aeruginosa, E. coli and Y. pestis can be noticed. Independent of the distinction between orthologs and paralogs (parameter f in Equ. 1 - phylogeny for f. 0 not shown) two diverse clusters are represented in the phylogenetic graph (Fig. 6 with almost identical topology: Cluster I consists of pathways *ec00*, *ng00*, *ng01*, *nm00*, *nm01*, *pa00*, *pa01* and *yp00*.

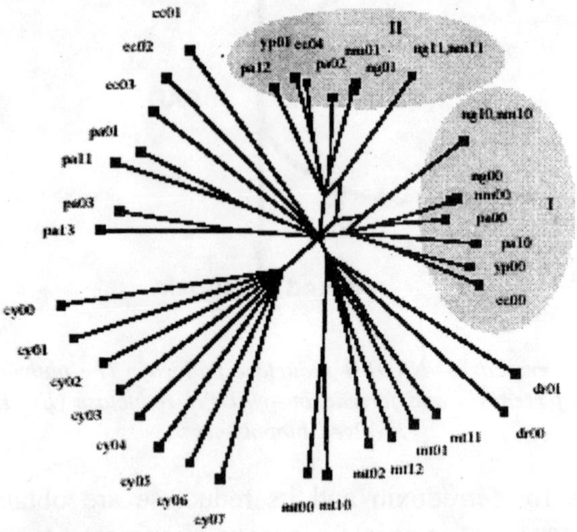

Figure 6. Ferredoxin – NADPH reductase pathway:Ph ylogenetic graph of the pathway drawn by SPLITSTREE2 with parameters f = 0.9, p = 2.3, and t = 1. Cluster I and II are referred to in the text.

Cluster II includes pathways *ec04*, *ng01*, *ng11*, *nm01*, *nm11*, *pa02*, *pa12* and *yp01*. Neisseria (*ngXX, nmXX*) are very close related and always show up as pairs in phylogenies. Quite surprising is the close similarity of *E. coli* (*ecXX*) to *Y.* pestis (*ypXX*). Studying corresponding pairs between cluster I and II indicates diversity in ferredoxins rather that in reductases. This gives hint to a higher demand of utilising ferredoxins for different tasks by nature than ferredoxin-NADPH reductases. A similar scenario can be observed in the case of the *mtXX* cluster. Pathways *mt01*, *mt11* and *mt02*, *mt12* cluster and differ significantly from *mt00*, *mt10*. This might gives hint to a specialisation of the pathway within the organism.

Pathways with Ferredoxin as Functional Role

Out of these approx. 50 pathways where ferredoxin plays a significant functional role 7 pathways are chosen for further investigations - the remaining non-used pathways are either assigned in one organism only or are absent at all due to missing functional roles in all organisms:

2-Oxoglutarate, Glutamine-Glutamate Anabolism (Reduced Ferredoxin)

NADPH–Oxidised Ferredoxin Electron Transport

NADPH–Oxidised Ferredoxin Electron Transport (Plasma Membrane)

H_2–H^+ Catabolism (Oxidised Ferredoxin)

H^+–H_2 Anabolism (Reduced Ferredoxin) (Plasma Membrane)

Nitrate–NH^+_4, OH^- Catabolism (Ferrocytochrome 'c_{552}',Reduced Ferredoxin) (Plasma Membrane, Cytosol)

Phosphoadenylylsulfate-Sulphide Anabolism

Table 3.List of Codes referring to ORF-names with corresponding Functional Roles

Code	Reductase	Ferredoxin
cy00	RCY25830	RCY45187
cy01	RCY25830	RCY08865
cy02	RCY25830	RCY02872
cy03	RCY25830	RCY15530
cy04	RCY25830	RCY13929
cy05	RCY25830	RCY13110
cy06	RCY25830	RCY46991
cy07	RCY25830	RCY42273
dr00	RDR02099	RDR01803
dr01	RDR02099	RDR01783
ec00	REC06248	REC02502

ec01	REC06248	REC06625
ec02	REC06248	REC01352
ec03	REC06248	REC00044
ec04	REC06248	REC05525
mt00	RMT05972	RMT02674
mt01	RMT05972	RMT00975
mt02	RMT05972	RMT04402
mt10	RMT00705	RMT02674
mt11	RMT00705	RMT00975
mt12	RMT00705	RMT04402
ng00	RNG00591	RNG01106
ng01	RNG00591	RNG00533
ng10	RNG00984	RNG01106
ng11	RNG00984	RNG00533
nm00	RNM01731	RNM00363
nm01	RNM01731	RNM00662
nm10	RNM00963	RNM00363
nm11	RNM00963	RNM00662
pa00	RPA07749	RPA01015
pa01	RPA07749	RPA08046
pa02	RPA07749	RPA01568
pa03	RPA07749	RPA07726
pa10	RPA05251	RPA01015
pa11	RPA05251	RPA08046
pa12	RPA05251	RPA01568
pa13	RPA05251	RPA07726
yp00	RYP02807	RYP00405
yp01	RYP02807	RYP01051

Fig. 7 shows a phylogenetic graph of 418 representations of these 8 pathways with a confidence level of $t = 0.5$. Thus 50% of functional roles have to be present. Otherwise the distinct pathway is not considered. Pathways of the same organism are found in one cluster, although a detailed analysis discloses a richer relationship. Pathways of E. coli and Y. pestis are close related to each other.

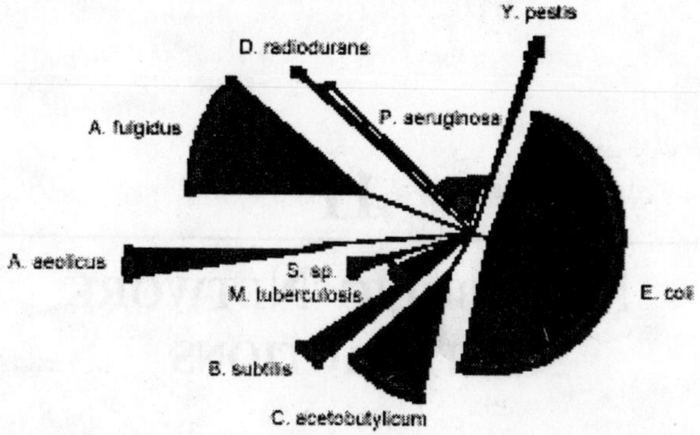

Figure 7. Ferredoxin-related pathways: 418 different representations of ferredoxin-related pathways are shown:10 for A. aeolicus, 48 for A. fulgidus, 10 for B. subtilis, 42 for C. acetobutylicum, 4 for D. radiodurans, 200 for E. coli, 12 for M. tuberculosis, 72 for P. aeruginosa, 16 for Synechocystis sp., and 4 for Y. pestis. The distance matrix is created with parameters f = 1, p = 1, and t = 0.5 and drawn by the program SPLITSTREE2.

In the same "super-cluster" but weaker related to the former species a second cluster is formed by M. tuberculosis, P. aeruginosa and Synechocystis sp.. For the remaining organisms (A. aeolicus, A. fulgidus, B. subtilis, C. acetobutylicum, and D. radiodurans) the significance of similarity is speculative.

11

METABOLIC NETWORK INTERACTIONS

Signaling and regulatory pathways consist of some number of components—such as genes, proteins and small molecules—wired together in a complex network of intermolecular interactions. As systems biology emerges in the post-genomic era, the emphasis is shifting from annotation of individual genes and gene products to ascertaining how DNA-protein and protein-protein interactions occur within a complex network of structural, metabolic, and regulatory pathways in cells.

This goal is, of course, aligned with that pursued in "reductionist" molecular cell biology for the past two decades, in which efforts to identify and characterise pathways typically have proceeded in a component-centric manner, beginning with an initial gene or protein of particular interest and attempting to ascertain other genes and/or proteins involved in the same pathway.

However, although component-centric approaches have been successful in assembling most of the available knowledge about pathways to date, they have several inherent difficulties. First is the time required: accurate models of pathway function emerge only after evidence is accumulated over many years, with the work of many researchers at many laboratories.

Second, these approaches do not directly reveal how multiple pathways influence each other or reveal this crosstalk only accidentally. Third, the vast amount of information on the various intracellular pathways remains fairly decentralised, buried across primary literature or within narrowly defined reviews.

Systems biology can offer an accelerated approach to this goal of identifying networks by which genes and proteins interact to carry out cellular operational and regulatory functions, using computational mining methods on high-throughput experimental data. The resulting high-level models are finding increasing utilization as tools for drug discovery, both by small companies as well as large pharmaceutical companies such as Eli Lilly and Novartis. In general, the resulting "high-level," topological models can then serve as a foundational prelude, if one wishes, for more familiar "low-level," mechanistic models.

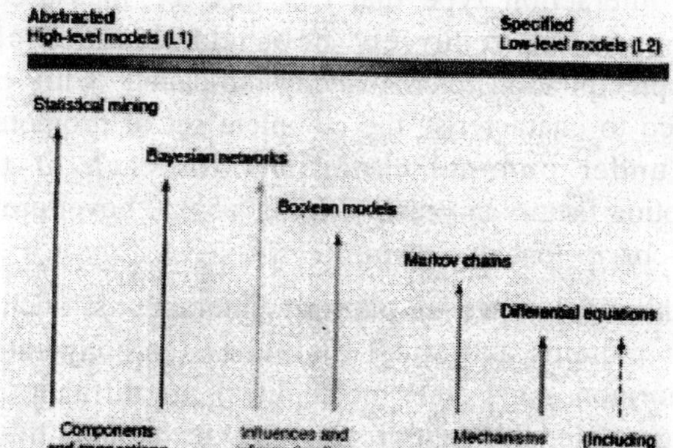

Figure 1. A diverse spectrum of high-to-low modelling approaches.

This technological developments are enabling to define and interrogate these pathways more directly and systematically than ever before, using two complementary approaches. First, it is now possible to systematically measure the molecular interactions themselves, by screening for protein-protein, protein-DNA and small molecular interactions. Several methods are available for measuring protein-protein interactions on a large scale-two of the most popular being the yeast two-hybrid system and protein co-immunoprecipitation (coIP) in conjunction with tandem mass spectrometry.

Although the vast majority of protein interactions have been generated for the budding yeast Saccharomyces cerevisiae, protein interactions are becoming available for a variety of other species including Helicobacter pylori and Caenorhabditis elegans and are catalogued in public databases such as BIND and DIP™ A current drawback of these high-throughput measurements is an associated high error rate. One approach for addressing this problem may be to integrate several complementary data sets together to reinforce the common signal.

Protein-DNA interactions, as commonly occur between transcription factors and their DNA binding sites, constitute another interaction type that can now be measured at high throughput. The relatively new technique of chromatin immunoprecipitation microarraying (so-called "ChIP-chip") has been used to characterise the complete set of promoter regions bound under nominal conditions for each of the >100 transcription factors in yeast, yielding >5,000 novel protein-DNA interactions in that organism.

Additional types of pathway interactions, such as those between proteins and small molecules (carbohydrates, lipids, drugs, hormones and other metabolites), are difficult to measure on a large scale, although protein array technology might enable high-throughput measurement of protein-small molecule interactions in the near future.

In addition to characterising molecular interactions, a second major approach for interrogating pathways is to systematically measure the molecular and cellular states induced by the interaction wiring. For example, global changes in gene expression are measured with DNA microarrays, whereas changes in protein abundance, protein phosphorylation state, and metabolite concentrations can be quantified with mass spectrometry, Nuclear magnetic resonance (NMR) and other advanced techniques.

Of these approaches, measurements made by DNA microarrays are currently the most comprehensive (every mRNA species is detected); high-throughput (a single technician can assay multiple conditions per week); well characterized (experimental error is appreciable, but understood); and cost-effective (whole-genome microarrays are purchased commercially for US $50 to $1000, depending on the organism).

However, continued advances in protein labeling and separation technology are making the measurement of protein abundance and phosphorylation state almost as feasible, with the primary barrier being the expense and expertise required to set up and manage a mass spectrometry facility. Measurement of metabolite concentrations, an endeavor otherwise known as metabonomics, is currently limited not by detection (thousands of peaks, each representing a different molecular species, are found in a typical NMR spectrum) but by identification (matching each peak with a chemical structure is difficult).

Clearly, measuring changes in cellular state at the protein and metabolic levels will be crucial if researchers are to gain insight into not only regulatory pathways, but also those pertaining to the cell's signaling and metabolic circuitry. To arrive at a high-level topological model of a cellular network of interest, data on molecular interactions and states can be integrated in a multi-tiered strategy. First, the global molecular interaction scaffold is constructed from systematic measurements of protein-protein interactions, protein-DNA interactions and/or metabolic reactions.

In the case of budding yeast, a minimal set might include 14,941 protein-protein interactions catalogued in the DIP™ database; 5,631 protein-DNA interactions from a combination of TRANSFAC® and ChIP-chip; and 599 enzymatic reactions in MetaCyc. Second, this scaffold is filtered against changes in mRNA expression, protein expression and/or posttranslational modifications recorded in response to different cellular perturbations.

Networks within the interaction scaffold with mRNA or protein states that are significantly activated by perturbation are identified and mapped according to a computational search engine. The interaction pathways and complexes comprising the scaffold constitute topological models, which are then prime candidates for further verification and modelling as important signaling and compensatory mechanisms controlling the cellular perturbation response.

The key advance of these searches is that by integrating two complementary global data sets, it is possible to condense and partition the enormous quantity of data into a small number of relevant pieces suitable for lower level modelling. Examples of this general scheme have been reported in recent literature. Several groups have applied "coclustering" approaches to identify groups of proteins that are both differentially expressed under similar sets of conditions and closely connected by the same network of interactions in the scaffold.

In many cases, these "expression-activated networks" correspond to well known protein complexes, regulatory pathways or metabolic reaction pathways. Other groups have used probabilistic approaches to match changes in gene expression with the transcription factors that are most likely to regulate them directly. These methods start with a cluster of differentially expressed genes and incrementally choose a small set of transcription factors that, by virtue of their levels and/or interactions in the scaffold, can maximally predict the observed levels of differential expression in the cluster.

New transcription factors are added only if they lead to a sufficient increase in predictive power over the transcription factors already in the model. Several software tools are now available for visualising interaction scaffolds. For instance, the Cytoscape framework provides network visualisation, layout and annotation, as well as clustering of the network against expression data to generate topological network models. The PathwayTools component of the MetaCyc metabolic pathway database can superimpose enzyme expression levels on the map of biochemical reactions for a species, giving a good indication of which reaction pathways are most affected over a panel of growth conditions profiled by microarray.

Because DNA microarray technology is currently much more widespread than technologies for protein or metabolite profiling, the vast majority of these approaches have used gene expression profiling as the primary state measurement. Of course, pathway mapping methods based on mRNA profiling alone capture just one facet of a much larger and complex cellular response. As it becomes possible to measure cellular state at the protein and small-molecule level, researchers expect that algorithms similar to those described above will emerge.

Currently, omitting this information from the analysis means that regulatory networks not purely transcriptional in nature remained to be elucidated in this high-throughput computation-aided manner. Indeed, the next obvious challenge beckoning is to push forward from the current focus on transcriptional regulation to regulation occurring in the post-transcriptional and post-translational arenas, and especially arising from extracellular cues.

Network Inference and Models

In the area of network inference, the models are primarily static interconnection descriptions of collections of proteins, metabolites, and/or genes. The "inference" problem involves the estimation of the interactions of elements in the network, given (possibly time series) data of activities of different nodes (e.g., gene interactions from gene expression data). The goals of the inference problem are multiple, and include:

— hypothesis generation,

— design of experiment,

— understanding of cellular function, and

— unraveling design principles, among others.

The sources of information for these inference problems include large-scale deletion projects, and vast numbers of microarray experiments. In the early years of bioinformatics studies, the structural localisation properties were inferred (e.g., which transcription factors regulate the transcription of which genes), although experimental methods now exist for identifying protein-DNA interactions on a genomic scale, such as ChiP assays, that yield structural knowledge.

The mathematical structures invoked to capture network interactions are numerous, and include:

— Boolean Networks-in which the network is represented as a graph of nodes, with directed edges between nodes and a function for each node

— Petri Nets-another graph theoretic structure in which nodes (or places) are connected by arcs and activities are modeled by transitions

— Bayesian Nets-combine directed acyclic graphs with a conditional distribution for each random variable (vertices in graph)

— Signed Directed Graphs-another graph theoretic structure in which a signed directed edge is used to represent activation versus inhibition (depending on sign)

— S-systems-notably a dynamic approach in which polynomial nonlinear dynamic nodes are used to capture network behaviour

A significant challenge in constructing these network models from data, particularly for gene network models, is the fact that the node dimension (number of genes) can be on the order of 10,000-leading to a computationally untenable problem for inference (i.e., determination of 108 coefficients of interaction!). In reality the network is tremendously sparse and highly structured, such that there are orders of magnitude fewer "interactions" that must be captured with coefficients.

The knowledge that not every gene regulates every other gene, and the fact that not every transcription factor regulates every gene can be exploited to prune significantly the number of coefficients for network identification. A related concept that can be exploited is the knowledge that the low dimensional connection structures in these networks obey regular hierarchies, which create opportunities for structured model identification.

Many biophysical networks can be decomposed into modular components that recur across and within given organisms. One hierarchical classification is to label the top level as a network, which is comprised of interacting regulatory motifs consisting of groups of 2-4 genes. At the motif level, one can use pattern searching techniques to determine the frequency of occurrence of these simple motifs, leading to the postulation that these are basic building blocks in biological networks.

Many of these components have direct analogs in system engineering architectures. Consider the three dominant network motifs found in E. coli:

— feed-forward loop,

— single input module, and

— densely overlapping regulon.

Similar studies in a completely different organism, S. cerevisiae, yielded six related or overlapping network motifs:

— autoregulatory motif,

— feed-forward loop,

— multi-component loop,

— regulator chain,

— single input module, and

— multi-input module.

Beyond structural classification, one can analyse these motifs for their functional character, one finds the recurring dynamic functional motifs in circuits and signal processing:

— switches,
— oscillators,
— amplitude filters,
— bandpass filters,
— memory,
— noise filters, and
— noise amplifiers.

In effect, in both eukaryotic and prokaryotic systems, cell function is controlled by sophisticated networks of control loops that are cascading onto, and interconnected with, other (transcriptional) control loops. The noteworthy insight here is that the complex networks that underlie biological regulation appear to be constructed of elementary systems components, not unlike a digital circuit.

This creates opportunity for the network inference methods that incorporate such knowledge via constrained search methods, or exploiting prior knowledge in Bayesian frameworks. In addition to the two classes of models mentioned previously (based on complexity and detail), there is an intermediate class consisting of optimisation-based models. In many respects, this class has a hybrid character of empiricism and fundamental details.

The underlying assumption is that cells have been organised over evolutionary time scales to optimise their operations in a manner consistent with mathematical principles of optimality. The cybernetic approach developed by Ramkrishna and co-workers is founded on a simple principle; evolution has programmed or conditioned biological systems to optimally achieve physiological objectives.

This straightforward concept can be translated into a set of optimal resource allocation problems that are solved at every time step in parallel with the model mass balances (basic

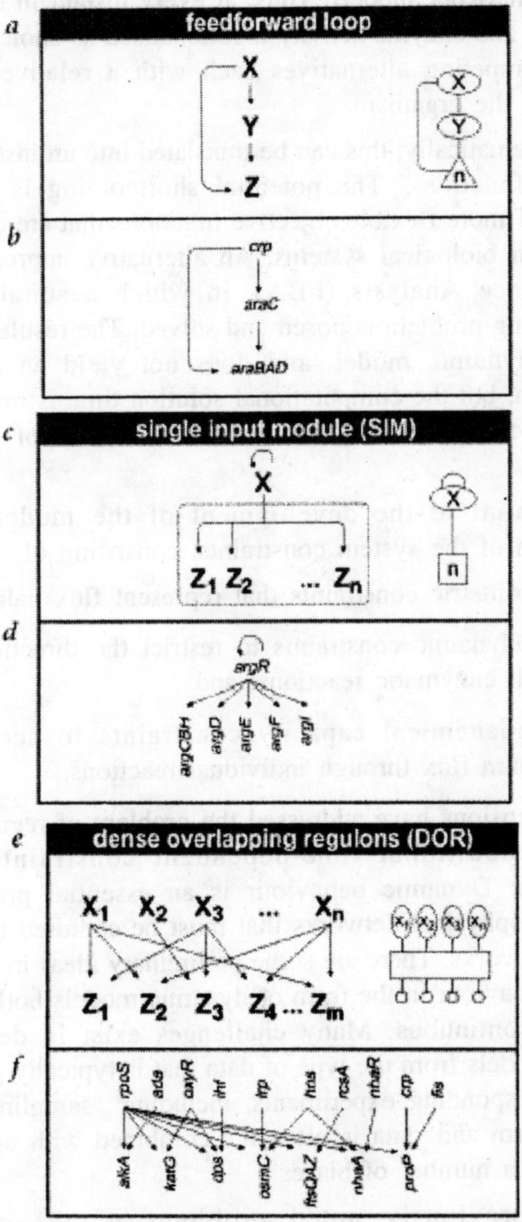

Figure 3. Network motifs found in the E.Coli transcriptional regulation network

metabolic network model). Thus, at every instant in time, gene expression and enzyme activity is rationalised as choice between sets of competing alternatives each with a relative cost and benefit for the organism.

Mathematically, this can be translated into an instantaneous objective function. The potential shortcoming is a limited handling of more flexible objective functions that are commonly observed in biological systems. An alternative approach is the Flux Balance Analysis (FBA), in which a suitable linear programming problem is posed and solved. The resulting model is not a dynamic model, and does not yield an analytical formulation, but the computational solution time is modest, and the approach has yielded success for a number of biological examples.

Essential to the development of the model are the formulation of the system constraints, consisting of:

— stoichiometric constraints that represent flux balances;

— thermodynamic constraints to restrict the directional flow through enzymatic reactions; and

— physicochemical capacity constraints to account for maximum flux through individual reactions.

Recent extensions have addressed the problem of regulation by including additional time-dependent constraints in the formulation. Dynamic behaviour is an essential property of complex biophysical networks that must be captured in models of those networks. There are some preliminary ideas in capturing network behaviour in the form of dynamic models-both discrete time and continuous. Many challenges exist in developing dynamic models from the type of data that is typically generated in the corresponding experiments, including: sampling rate is rarely uniform and data is often the combined with other labs, introducing a number of biases.

The previously noted problems of the curse of dimensionality are more pronounced in the case of dynamic models, if one augments the network interconnection dimensionality with a large number of possible dynamic states (activated, repressed, silenced, etc.), let alone the full continuum of dynamic response.

VALIDATION, ITERATION, DISCRIMINATION, AND IDENTIFIABILITY

One of the major issues in the reverse engineering of a genetic regulatory network is the challenge of uniquely identifying the gene interactions (i.e., model parameters) from experimental data, such as gene expression profiling. This issue, known as identifiability in control theory, deals with the information content of the data; the quantity and quality of the measurements with respect to the model parameters.

Typical knockouts involve so-called "direct effects" in which the expression level of various genes is altered in a network arrangement that involves direct connectivity to cis-regulatory elements of downstream genes (possible multiple cascades). An "indirect effect" can also be used in which a mediating component (e.g., mRNA) is introduced to correct an intermediate element in the direct action cascade described previously.

Coupled to this, noise in the measurements and the inherent stochastic nature of gene expression make practical identification of genetic regulatory networks difficult. In practice, the reverse engineering of a gene network should involve a careful design of the experiments using prior knowledge of the system, to obtain the most informative measurements. Further, this process should be iterative in which the result from each trial is used to better design the next experiment. Here, a measure of informativeness of data, such as the Fisher Information Matrix (FIM), can lead to a formal procedure for the optimal design of the experiment.

Aside from the aspect of the quality of data, another practical limitation in most (if not all) of the reverseengineering of a gene network is the limited quantity of data, in terms of sampling frequency and number of independent measurements. For example, although gene expression profiling can provide high-throughput data to estimate interactions among thousand of genes, this method still does not depict the protein-mediated regulatory effects.

In many cases, parameter estimation from limited measurements suffers from stringent computational requirement and degeneracy, where many parameter combinations give similar agreement to the observed behaviour. Here, measurement selection procedures can help identify the combination of measurements that give the best identifiability.

Given the iterative nature of this framework for model development and refinement of experimental protocols, a termination criterion must be established. In the application domain of systems engineering, it is understood that for certain experimental data, it is not possible to confirm whether the model is really valid; however, one can conclude whether the model is not contradicted by the given data.

Such model (in)validation tests can be formulated for the network inference problems described in this chapter, and are usually based on the difference between the simulated and measured output and some statistics about these differences. Typical statistics for the model errors include maximum absolute value, mean value and variance. These methods are slowly migrating from the engineering domain and are likely to find greater application in systems biology as experimental methods are refined, and closer collaborations are developed between modelers and experimentalists.

Regression

Most of the network inference work undertaken to date has been aimed at elucidating relationships among components in a network. However, the operation of a network is generally important only within the context of a physiological function it carries out or regulates at the level of individual cells or cell populations (e.g., tissues). Thus, it is essential to consider efforts to elucidate relationships between network components (or, more appropriately, network component properties such as levels, states, locations, and/or activities) and downstream cellular behavioural functions.

Computational models for these kinds of "signal-response" relationships are much more difficult to formulate than for the more commonly-studied "cue-signal" relationships by which network component properties are governed by extracellular (or intracellular) stimuli, because the biochemical and biophysical processes involved in the former are much less well understood as well as less proximal. Nonetheless, several research efforts are underway that take as their objective the development of inferring "signal-response" relationships for network component regulation of cell functions.

One class of methods which appear to be useful for inferring dependence of cell functions on network component properties are those founded on principal components analysis, including a variant termed network components analysis. This class of methods permits determination of the most critical combinations of network component properties for correlation, and even prediction, of functional responses. A second class of methods also being employed for this purpose is decision tree analysis.

An advantage of decision tree analysis is that the combinations of network component properties associated with functional responses are explicitly delineated in direct manner, permitting the "logic" of how the network component properties combine to govern functional behaviour to be viewed and interpreted easily. Finally, Bayesian network analysis has been used for a similar purpose. Here, the "logic" connecting network component properties to one another and to the functional behavioural response is more complex to interpret, but is nonetheless available.

12

MICROBIAL METABOLIC NETWORKS

Systems biology aims at an understanding of the genotype-phenotype relations brought about by cellular networks. Mathematical models as formal representations are central for handling the associated complexity. Recently, model-based analysis of microorganisms has begun, for instance, to reveal functional modules in metabolic and transcriptional networks, to predict cellular behaviour from genome-scale physicochemical constraints, and to suggest novel design principles for well-studied bacterial subsystems such as chemotaxis.

The unprecedented increase in the number of sequenced genomes—many of which are microbial—and the corresponding rise of genomics, proteomics, and other 'omics' have transformed the 'operation mode' of biology. Comprehensive measurement technologies yield largescale datasets on many cellular components and their interactions. More generally, this fuels the interest in the operation of cells as integrated systems or networks, complementing the focus on individual components that prevails in molecular biology. Particularly for microbial cells, the complete determination of cellular networks is a realistic perspective.

The analyses of proteome, the ensemble of proteins associated with a given cell, organelle or other sub-cellular component at a given physiological state of cells have been made possible by the sequencing of genome of a number of organisms.

The proteins bring about the functions of a cell, the study of their ensemble would help better understand the function of cell and also enable integration of the function at molecular level. Understanding the proteome of microorganisms provides two broad advantages, one is the utilisation of microorganisms or biotechnological exploitation, that is, for the production of metabolites and the other is to deterring the pathogenic and parasitic microbes by understanding host-pathogen interactions or interaction of the microorganisms with multi-cellular systems.

The integration of message transfer at different levels within the cells is not completely understood and analyses of proteome may help improve that understanding, which is expected to improve metabolite production and also better targeting of molecular interactions as therapeutic strategy. Biotechnology is the exploitation of the living organisms either directly or using derivatives of them for the benefit of humankind. The derivatives are often exploited is different ways compared to their counter natural processes within the cells. The chemical or biochemical engineering processes operate with purified compounds (substrates or reactants and catalysts or enzymes).

The processes with purified catalysts usually turn the substrates into product(s). The engineering processes are difficult to maintain at a steady state of production, require very careful maintenance of the operation variables. The presence of impurities and variation in any parameters results in dramatic decrease in productivity despite employing the same cellular catalyst in engineered processes. A cell lack perfect compartmentalisation, although a definite degree of separation or compartmentalisation exists within the cells and produces a large number of chemicals at any given time in a mixed state.

In contrast to engineered systems which has a great deal of emphasis on purity of reagents, lack robustness and not rapid in response to perturbation in operation variables, all cells demonstrates a great degree of tolerance to impurity, are impervious to disturbances and mostly have a rapid response to an external stimuli. The perturbations in operation variables are often detrimental for chemical and bioengineered systems. Adapting control features in engineered processes from cells would enable better process control and may enable designing methods for new chemicals employing cell derived catalysts.

The cell can be viewed as ensemble of network of interactions such as protein-protein, protein-nucleic acids and protein-small molecule. Several protein and small molecules constitute anabolic, catabolic and amphibolic pathways which are often interconnected and referred to as metabolic pathways. Inside the cells, specific metabolic networks are driven by demand for robustness to uncertain environments and use often imprecise components, yet in response to an external stimuli it produces the precise yield of chemicals necessary, with a great deal of molecular understanding these paradoxical features remains to be understood even today.

It has been put forth that these paradoxical features are neither accidental nor artificial but derive from a deep and necessary interplay between complexity and robustness, modularity, feedback and fragility. In anabolic pathways a substrate enters into the pathway and acted upon by a number of proteins in series for a particular conversion, which is necessary for a catalytic reaction involving another substrate received from external cellular environment. All of this occurs in presence of entities unrelated to pathway.

The importance of metabolic networks within the cell has been realised and they have been found interesting and complex enough to be the subject of intense mathematical and computational analysis. Thousands of components of a living cell are interconnected and are in the dynamic state. Thus the functional properties of the cells are encoded by a set of complex intracellular interactions at molecular level. The architecture of metabolic networks inside the cell has been found to have a highly heterogeneous scale-free topology. This structure of network also allows simultaneous tolerance to random errors and determines the key aspects of functionality and regulation.

Protein Interactions

A number for high-throughput methods for detecting protein interactions especially for microbial systems are now available. Protein-protein interactions play crucial roles in the execution of various biological functions. Interactions with other proteins are of extreme importance and can serve as highly informative hints for functional prediction: physical association between a novel protein and a well characterised one readily indicates that the former has a function related to that of the latter. Comprehensive analyses of protein-protein interactions thus constitute an integral part of functional genomics.

The natures of proteins are far more complex than nucleic acids. It is apparent that a variety of methods would be needed to determine genome wide protein-protein interactions for any organism or even for subcellular organelles. These approaches can be divided into two categories: the top-down proteomic approach and the bottom-up genomic ones. The proteomic approaches are represented by the mass spectrometric and protein chip analyses of native protein complexes. The genomic approaches are where protein encoded in the genome of interest is expressed to evaluate interactions. These include the techniques such as yeast two-hybrid system and phage display.

Purifications of Protein Complexes

The purifications of protein complexes are achieved by immunoprecipitation using column of beads containing coupled antibodies to one of the components of complex. Often a variation of method is utilised where either a bait protein or tagged components are used. The constituent protein components of the complex is then separated by gel electrophoresis and subjected to analyses by mass spectrometry. Usually the excised gel bands (1-D gels) or spots (2-D gels) of proteins are subjected to in-gel proteolytic digestion (usually trypsin) and subjected to mass spectrometric identification. Two techniques are commonly used the first, Tandem affinity purification (TAP) allows rapid purification under native conditions of complexes, even when expressed at their natural level.

The TAP method requires fusion of the TAP tag, either N- or C-terminally, to the target protein of interest. The TAP method was initially developed in yeast but can be successfully adapted to various organisms. Its simplicity, high yield, and wide applicability make the TAP method a very useful procedure for protein purification and proteome exploration. The second technique is, high-throughput mass spectrometric protein complex identification (HMS-PCI) protocol where a number of predicted interacting partner proteins are used as baits and associated proteins are identified. This method was also initially developed for yeasts.

Numerous protein complexes have been identified using this method including several new interactions in a number of signaling pathways and in the DNA damage response. The TAP approach has the advantage of working with the protein in its native state within the cells and the tagged proteins provides avenue to determine internal consistency. However, it has the disadvantage of missing the weak binding partners or weak association complexes and may not detect complex partners that associate with protein that undergoes transient modification subject to irreversibility on changing the conditions. The HMS-PCI approach may result in formation of some unnatural complexes and render detection of some partners impossible due to blockage of sites on protein as a result of association with the bait present in excess amount than in native state.

Protein Chip

Protein chips coated with antibodies or chemicals help capture the protein or biomarkers. The captured proteins or biomarkers can be identified by using surface enhanced laser desorption/ ionization (SELDI) mass spectrometry. The technique of SELDI is also referred as Protein Chip Technology. Currently protein biochips can be utilised to profile and compare protein expression in normal and diseased states particularly in the area of cancer, infectious disease and toxicology. The technique has the limitation at level of capture and the ability of ionization of low abundance proteins.

Yeast Two-hybrid Assay

Pairs of proteins to be tested for interaction are expressed as fusion proteins (or hybrids) in yeast with one protein being fused to a DNA-binding domain (bait) the other to a transcriptional activator domain (prey). The plasmids carrying the DNA sequences coding for the bait and prey proteins are introduced into the same yeast cell using a mating strategy. The system is designed so that the bait and prey proteins interactions leads to transcriptional activation resulting in expression of a reporter gene that allows detection of the interaction. This technique enables detection of transient and unstable interactions and also in vivo interactions.

However, this technique allows only pair-wise testing proteins and interaction of multiple proteins at a given time cannot be investigated. The investigation of proteins resident of sub-cellular organelles by this technique deviates from their native physiological setting and often provides results that are not true.

Correlated mRNA Expression

The genes showing similar transcriptional response under different cellular conditions are grouped by systematic measurement of the mRNA levels. Usually such gene groups consist of a large number of genes that codes for physically interacting proteins. In eukaryotes, highly coordinated expression of genes functioning in common processes is a widespread phenomenon. These sets of co-regulated genes ('synexpression groups') may be the key determinant facilitating evolutionary changes that lead to species diversity. The development of DNA micro-arrays for the simultaneous monitoring of thousands of transcripts has allowed global insights into gene expression.

The technique of synexpression has helped determine a number of interacting proteins for yeast cell-cycle analysis and potential drug targets in rat brain by EST analyses. The results indicate the usefulness of synexpression for exploring the functional genomics for information forthcoming from various genome projects. This is an indirect but in vivo technique and may cover a wide range of cellular conditions than other methods, but is a relatively inaccurate predictor of direct physical interaction.

GENETIC INTERACTION

Two mutations are synthetically lethal if cells with either of the single mutations are viable but cells with both mutations are nonviable. Usually these genes are functionally associated and their protein products undergo physical interaction. Such genetic interactions are being studied in an all-versus-all approach in yeast. As with the suppressor analysis, synthetic lethal mutations often indicate that the two mutations affect a single function or pathway. Synthetic lethal phenotypes are diagnostic of an interaction between the products of the two mutant genes in the cell. Combinations of synthetic lethal mutants can also give information about what products are needed to complete a cellular process. This is also an indirect in vivo technique, amenable to unbiased genome-wide screening for protein-protein interaction.

Phage Display

The phage display approach is a powerful tool, which was initially developed to detect antibodies. Phage carrying V genes were displayed on the surface of fd bacteriophage which binds specifically to antigen and can be isolated by affinity chromatography. The use of the technique has expanded and phage libraries displaying random peptides can be used to select for ligands able to bind specific target molecules, to mimic non-proteinaceous ligands and also as a tool to map epitopes recognised by the antibodies. The display of the proteins or their functional domains provides a system for the analysis of structure-function relationships, novel peptide vaccines and others. The technique suffers from limitation where interaction depends on conformation or needs correct folding for binding, which is not conferred by expressed truncated peptides.

Three types of evidence may be found in a genome wide scan indicating interacting partners or potential interaction among the proteins in question, the first type of evidence being, the proteins that are interaction partners are often encoded by conserved operons in prokaryotes. The second type of evidence being, since interacting proteins interact together to provide a function or absence of a partner may impair the others and therefore in the genome wide scan the interacting proteins are likely to be either present or absent together and unlikely to be found alone in absence of other key partners.

The interacting proteins are likely to have a similar 'phylogenetic profile'. The third type of evidence being, the presence of fusion of seemingly unrelated proteins into a single polypeptide chain suggests a physical interaction. The in silico techniques are rapid, relatively inexpensive and with the rapid sequencing of genomes their utility is ever expanding. However, in silico experiments requires assigning orthology relationship between proteins, and fails in prediction where such relationship cannot be established and this has led their application remaining confined mainly on prokaryotes.

Metabolic and Protein-protein Interaction Network

The cellular processes in a microorganism are seamlessly integrated through a complex network of constituent components and reactions. Systematic comparative mathematical analysis of the metabolic networks of several microorganisms have shown that, despite significant variation in their individual constituents and pathways, these metabolic networks have the same topological scaling properties and show striking similarities to the inherent organisation of complex non-biological systems. This indicates that organisation of metabolic networks is identical for all living organisms, and also has similar design principles such as being robust, error-tolerant and being scale-free.

The metabolic network entails large-scale organisation of interactions among all cellular constituents and probably has a common blueprint among all organisms. The robustness of various cellular processes is rooted in the dynamic interactions among its many constituents, such as proteins, DNA, RNA and small molecules. Scientific developments have improved our ability to identify the design principles that integrate these interactions into complex systems.

Discovery of genomes now allow development of organism-specific connectivity maps of metabolic and, to a lesser extent, other cellular networks, however, the large number and diversity of the constituents and reactions that form such networks, these maps are extremely complex, offering only limited insight into the organisational principles. The study of network properties has endowed ability to address in quantitative terms the structure of these cellular networks and has advanced our understanding of the generic properties of complex networks.

Properties of Network

Complex networks occurring in nature share common global statistical features including "small-world" property of short-paths between two nodes and highly clustered connections. In many such networks there are few nodes with many more connections than the average node connections and such networks have been termed as "scale-free". In scale-free networks the fraction of nodes having k edges, p(k), decays as power law $p(k) \sim k^{-y}$ (y is between 2 and 3). Understanding of basic structural elements particular to each class of network is necessary to define the networks properly. Patterns of interconnections between the nodes have been termed as "network motifs".

Understanding of the network motifs is expected to help uncover the structural design principle of the networks. Motifs are expected to define the universal classes of networks and unraveling them may help uncover the basic building blocks of most networks. The classical random network theory was introduced by Erdös-Rényi, based on the assumption that each pair of nodes (that is, constituents) in the network is connected randomly with probability p, leading to a statistically homogeneous network in which, despite the fundamental randomness of the model, most nodes have the same number of links, $<k>$.

In particular, the connectivity follows a Poisson distribution that peaks strongly at $<k>$ implying that the probability of finding a highly connected node decays exponentially ($P(k) \sim e^{-k}$ for $k >> <k>$). On the other hand, empirical studies on the structure of the World-Wide Web, Internet and social networks have shown serious deviations from this random structure, demonstrating that these systems are described by scale-free networks for which P(k) follows a power-law, $P(k) \sim k^{-\gamma}$. Unlike exponential networks, scale-free networks are extremely heterogeneous their topology being dominated by a few highly connected nodes (hubs), which link the rest of the less connected nodes to the system.

The distinction between scale-free and exponential networks emerges as a result of simple dynamical principles. Understanding the large-scale structure of cellular networks can provide not only valuable universal structural information, but could also lead to a better understanding of the dynamical processes that generated these networks. Any two nodes in the system can be connected by relatively short paths along existing links also known as "small-world character" is a general feature of many complex networks. In metabolic networks these paths correspond to the biochemical pathway connecting two substrates. The degree of interconnectivity of a metabolic network can be characterised by the network diameter, defined as the shortest biochemical pathway averaged over all pairs of substrates.

Power-law Connectivity

An important consequence of the power-law connectivity distribution is that a few hubs dominate the overall connectivity of the network. The sequential removal of the most connected nodes leads to sharp rise in the diameter of the network eventually disintegrating the network into isolated clusters that are no longer functional. Scale-free networks also demonstrate unexpected robustness against random errors. A unique feature of metabolic networks, as opposed to non-biological scale-free networks, is the apparent conservation of the network diameter in all-living organisms.

Within the special characteristics of living systems this attribute may represent an additional survival and growth advantage, as a larger diameter would attenuate the organism's ability to respond efficiently to external changes or internal errors. For example, if the concentration of a substrate were to suddenly diminish owing to a mutation in its main catalyzing enzyme, offsetting the changes would involve the activation of longer alternative biochemical pathways, and consequently the synthesis of more new enzymes, than within a metabolic network with a smaller diameter. An important known consequence of the inhomogeneous structure is the network's simultaneous tolerance to random errors, coupled with fragility against the removal of the most connected nodes.

The simultaneous emergence of an inhomogeneous structure in both metabolic and protein interaction networks suggests that there has been evolutionary selection of a common large-scale structure of biological networks and indicates that future systematic protein-protein interaction studies in other organisms will uncover an essentially identical protein-network topology. The correlation between the connectivity and indispensability of a given protein confirms that, despite the importance of individual biochemical function and genetic redundancy, the robustness against mutations in organisms is also derived from the organisation of interactions and the topological positions of individual proteins

An important feature of scale-free networks is the existence of a few highly connected nodes that participate in a very large number of metabolic reactions. With a large number of links, these hubs integrate all substrates into a single, integrated web and therefore the existence of fully separated modules is prohibited in highly integrated module-free metabolic network organisation. The analyses of the functional capabilities of metabolic networks by a number of approaches have indicated the existence of separable functional modules.

The metabolic network of many organisms (for example,E. coli) is known to possess a highclustering coefficient that suggests modular organisation. The apparent dichotomy of presence of modular organisation in contrast to scale-free nature of metabolic networks was reconciled by hierarchical organisation. Average clustering coefficient defined as $C_i = 2n/k_i(k_i - 1)$, was calculated for 43 organisms, where n denotes the number of direct links (distinct substrates present in the metabolic network) connecting the k_i nearest neighbors of node i, is equal to 1 for a node at the center of a fully interlinked cluster, and it is 0 for a metabolite that is part of a loosely connected group.

Therefore, C_i averaged over all nodes i of a metabolic network is a measure of the network's potential modularity. The average clustering coefficient was determined for 43 organisms and was found to be an order of magnitude larger than that expected for a scale-free network of similar size suggesting that metabolic networks in all organisms are characterised by a high intrinsic potential modularity. The clustering coefficient of metabolic networks is independent of their size. To resolve the apparent contradiction of existence of modularity and scale-free network a simple heuristic model of metabolic organisation was proposed and referred to as a "hierarchical" network.

The architecture of such a network integrates a scale-free topology with an inherent modular structure. It has a power law degree distribution with degree exponent $\gamma = 1 + (ln\ 4)/(ln\ 3) = 2.26$, in agreement with $\gamma = 2.2$ and has a clustering coefficient C is about 0. 6 (independent of the size of the network) is also comparable with that observed in metabolic networks. In deterministic scale-free networks, the clustering coefficient of a node with k links follows the scaling law $C(k) \sim k^{-1}$. This scaling law quantifies the coexistence of a hierarchy of nodes with different degrees of modularity.

The higher a node's connectivity, the smaller its clustering coefficient that asymptotically follows the $1/k$ law. The $C(k)$ function for the metabolic networks of 43 organisms was calculated to determine whether hierarchical organisation is present in cellular metabolism. For each organism, $C(k)$ was represented by $C(k) \sim k^{-1}$, in contrast to the k-independent $C(k)$ predicted by both the scale-free and modular networks providing evidence for an inherently hierarchical organisation. Hierarchical modularity reconciles all observed properties of metabolic networks: their scale-free topology; system size-independent clustering coefficient; and the power law scaling of $C(k)$.

The organisation of metabolic networks is likely to combine a capacity for rapid flux reorganisation with a dynamic integration with all other cellular function. The system-level structure of cellular metabolism is approximated by a hierarchical network organisation with seamlessly embedded modularity. In contrast to commonly assumed existence of a set of modules with a non-uniform size potentially separated from other modules, the metabolic networks has many highly integrated small modules that group into a few larger modules, that in turn is integrated into even larger modules.

The scale-free topology has been found at many different organisational levels, genetic protein interaction and protein domain networks within the biological networks. For biological systems, hierarchical modularity is consistent with the notion that evolution may act at many organisational levels simultaneously. The accumulation of many local changes, which affect the small, highly integrated modules, could slowly impact the properties of the larger, less integrated modules. The emergence of the hierarchical topology through copying and reusing existing modules and motifs reminds the results of gene duplication and offers a key role to the modules that appeared first.

Inferring Microbial Genetic Networks

Cellular processes are governed by extensive, interconnected networks of genes, proteins, and metabolites. An important challenge in systems biology is to determine the structures and mechanisms by which these complex networks control cell processes. Recent studies of cellular networks include methods for identifying gene and protein interactions, regulatory modules, and global structural properties. These methods are yielding valuable information and insights, but they often fail to identify the regulatory role of individual elements or the system-wide functional properties of a given network.

Computational modelling and simulation also help toward understand network functions, but their utility depends on amassing extensive information, including regulatory structures, network connections, rate constants, and biochemical concentrations. Such data are generally not available, particularly for larger regulatory networks, due to the high cost and slow speed of experimentation.

In recent work, some researchers developed an efficient method for inferring the basic structure and function of microbial genetic networks using limited experimental information. The method, called network identification by multiple regression (NIR), quantitates the influence of genes on one another using measurements of a microbe's transcriptional response to genetic perturbations. NIR assembles this information into a network model that can be used to interpret additional experimental data, make predictions about network behaviour, and identify useful control points in the regulatory network.

NIR-inferred network models will be of great value in a variety of applications, including bioremediation, bioproduction of chemicals, and development of new antibiotics. For instance, efforts to develop antibiotics based on natural products or on more recent genomicbased methods that focus on lethal or "essential" genes as drug targets have delivered few new antibiotics to the clinic. Using network maps of bacterial pathways could help to overcome the shortcomings of these and other traditional methods-in part, by identifying genetic pathways and key regulators that serve as targets for new antimicrobial agents that circumvent mechanisms of resistance.

The NIR method departs from structural approaches such as twohybrid and DNA-binding assays. Structural approaches define networks through measurements of probable physical interactions, but they typically do not provide information on functional relationships between genes. On the other hand, the NIR method, which is a form of system identification, identifies quantitative regulatory relationships between genes under observed cellular conditions. Thus, with the NIR method, we can analyse and predict dynamic responses of gene networks, thereby gathering information that complements the findings of structural approaches.

To apply the system identification approach to gene networks, controlled perturbations, such as a set of gene overexpressions, are delivered to a cell, global measurements of that cell's response are obtained, such as through transcript profiling, and an algorithm is applied to the data that identifies or learns a model of the genetic network (Figure 1A). This model helps to define the relationship between input variables such as RNA concentrations and output variables such as gene transcription rates (Figure 1B).

The challenge in applying a system identification method is to select a model, a learning algorithm, and an experimental design that accurately and efficiently define the network of interest. In the NIR method, they chose to represent gene interactions with a linear model. That is, the rate of transcription

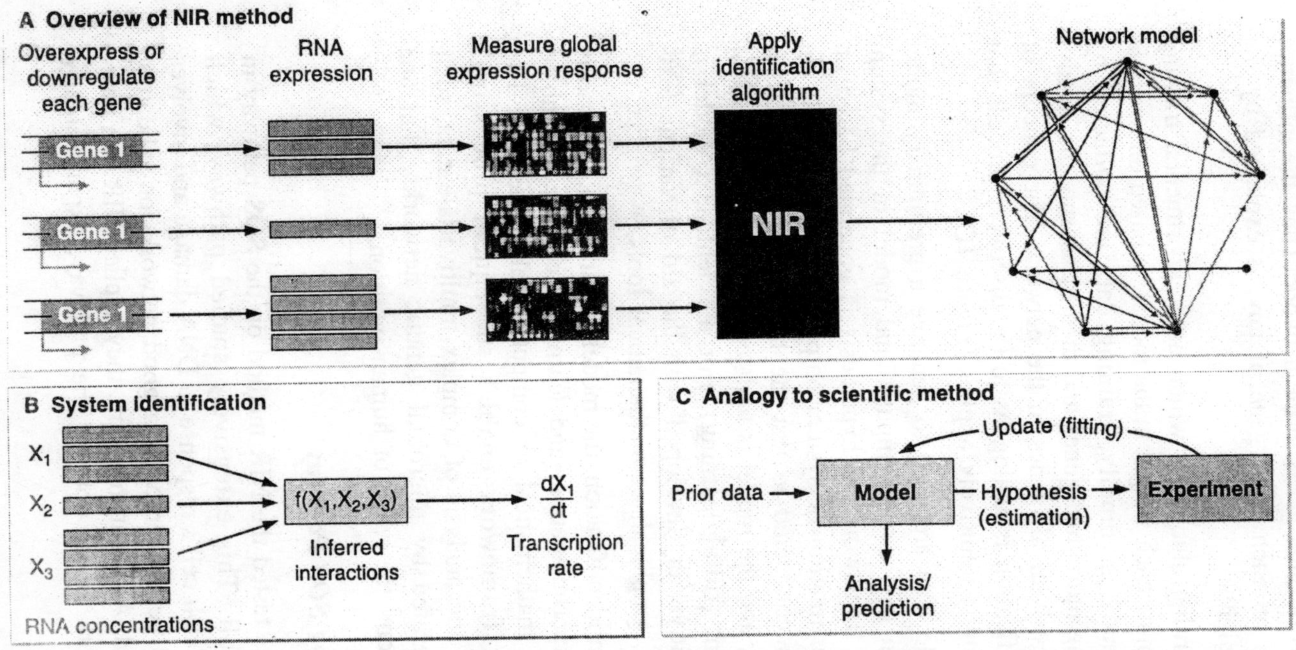

Figure 1. Overview of the NIR method. (A) A structured set of perturbations is delivered to cells, such as the overexpression or downregulation of one or more genes in each experiment. (B) In system identification, the inferred model relates a set of input variables to output responses using a mathematical function. In the NIR method, the input variables (X_1, X_2, X_3), are RNA concentrations of each gene, the output responses are transcription rates of each gene (dx_i/dt), and the function f is a weighted sum of the input variables. (C) System identification is analogous to the scientific method.

of each gene is represented as a weighted sum of the RNA concentrations of other genes in the network. Thus, the influence of each gene on each other gene, if any, is explained by the calculated weights.

The learning algorithm, which uses experimental training data to determine each weight in the model, is based on multiple regression analysis. Finally, training data are collected by overexpressing individual genes and then measuring the steady-state RNA levels of all genes in the network. The concept of system identification is analogous to the discovery process that a typical scientist intuitively applies (Figure 1C).

Available data are used to generate a preliminary model of the system (a formal computational model in the system identification framework), a hypothesis is generated based on that model (estimation of system outputs based on inputs), an experiment is conducted to test the hypothesis (collection of training data), and the model is updated based on the results of the experiment (fitting/learning). For a scientist, a model is typically intuitive, or developed with the aid of various text-based, logical, graphical, or mathematical tools.

In system identification, the model construction, hypothesis generation, experimentation, and learning steps are assembled into a formal mathematical or computational framework. The rigor of that framework enables subsequent processing, interpretation, and analysis of complex, multivariate data, such as those generated by microbial genetic networks, that are ordinarily beyond the reach of human intuition.

Testing NIR by SOS Pathway

The researchers tested the NIR method on the SOS pathway in Escherichia coli. This extensively studied pathway, which regulates an E. coli cell's response to DNA damage and involves more than 100 genes, serves as a good network for validating the NIR method. As a starting point, they applied NIR to a nine-gene subnetwork at the core of the pathway, experimentally altering an inducible plasmid to overexpress each of those genes and measuring RNA responses using realtime PCR.

With the NIR method, they correctly identified 25 already recognised regulatory relationships among those nine genes as well as 14 additional relationships that either are novel regulatory pathways or false-positive findings. Moreover, the NIR-determined network model correctly identified recA and lexA, the known principal regulators of the SOS response, as having the strongest influence, or largest regulatory weights, on other genes within this network. Thus, the model can be used to suggest which genes should be perturbed to elicit a particular response from the network-a capability of great value in optimising bacteria for environmental remediation or bioproduction of compounds.

They also used the NIR-determined network model together with additional experimental data to identify genes that mediate the network response to a drug or other stimulus. For example, they treated E. coli cells with mitomycin C (MMC), an antibiotic that leads to the formation of single-stranded DNA, thereby activating the RecA protein. Activated RecA subsequently significantly upregulates all genes in the test SOS network. When we applied this inferred network model to analyse experimental MMC response data, we found that it correctly identified recA as the key mediator of MMC bioactivity.

The model also identified the DNA-translesion repair polymerase, umuDC, as a second mediator, albeit at a lower significance level. They also applied the model to publicly accessible data that were obtained by using microarrays to assay responses of E. coli to various stimuli. Here again, the network model correctly identifies recA as the key mediator of the SOS response to DNA damage caused by UV irradiation and also treatments with the quinolone antibiotic pefloxacin. But recA is not identified as a mediator of E. coli response to novobiocin treatment, which does not damage DNA.

In addition, umuDC is identified as a mediator of the cellular response to DNA damage resulting from UV irradiation, but not to the DNA damage resulting from pefloxacin treatment. The absence of umuDC as a mediator of quinolone-induced DNA damage suggests that the genes involved in DNA repair following quinolone damage are different from those associated with damage resulting from MMC and UV irradiation.

One of the goals in using NIR, as with using most system biology methods, is to understand properties of cellular and biochemical systems that are not apparent from studying individual components. Some researchers presume that to understand such global properties, it is necessary to build extensive computational models that integrate most of the biochemical details of a cell or gene network. However, building such a model is unrealistic both computationally and experimentally because cells are too complex.

Moreover, for most practical purposes building such a comprehensive model is unnecessary. Traditionally, the complexity of cellular biochemistry is addressed by reducing a particular system to its components, which are then studied in isolation. But in following such an approach, valuable information about system properties is lost, precisely the opposite of what is desired in a systems analysis. Somehow, an alternative approach is needed in which the biochemistry is simplified in a way that preserves the information needed to describe system-wide properties of the network.

In system identification, that simplification is achieved by restricting the complexity of the model chosen to represent the system. For example, with NIR, they use a linear approximation of network interactions because it limits the number of required experiments, but still captures network properties of value in medical and biotechnological applications, namely by identifying major network regulators and mediators of chemical or environmental stresses.

However, this model may not be appropriate for analysing other behaviours of microorganisms, such as the genetic interactions that orchestrate bacteriophage infection. Other models of varying complexity, including Boolean, Bayesian, and neural network models, may be better suited for analysing such systems. In fact, one of the major challenges is to select an appropriate model structure that enables analysis of selected global properties while preserving computational tractability, speed, and experimental feasibility.

The NIR method has immediate applicability for improving antibiotics. For example, biofilms are involved in as many as 60% of human infections and are notoriously difficult to eradicate because cells in biofilms survive antibiotic doses several orders of magnitude higher than those sufficient to kill free-floating bacteria. Although mutations can lead to resistance against anti-infective drugs, genetic changes are not necessarily responsible for the increased antibiotic tolerance that occurs in biofilms. For instance, cells removed from biofilms often prove to be as susceptible to antibiotics as are their free- floating counterparts.

Moreover, reduced diffusion of drugs into biofilms does not generally account for their ability to withstand such drugs. For instance, fluoroquinolone antibiotics such as ofloxacin and ciprofloxacin readily penetrate biofilms and kill many cells. Yet, small numbers of cells within biofilms apparently survive regardless of the concentration of antibiotics applied, according to Kim Lewis and colleagues at Northeastern University in Boston, Mass. These "persistor" cells are believed to repopulate bio-films after antibiotic treatments cease, and thus to cause recurrent infections. In addition, such persistence may permit advantageous mutations to be amplified.

The mechanisms of persistence are not well understood. But persistence is likely a dynamic response of the cell orchestrated by multiple stress response pathways. A better understanding of these stress response networks, obtained using the NIR method, will be of great value in identifying productive targets for novel antibio-film compounds. The NIR-determined network model can also be used to identify genes that mediate the effects of a particular compound. Thus the network model could be of great value for optimising candidate antibiotic compounds, and could enable the development of novel classes of drugs that target the complex regulatory properties of genetic networks.

Identifying and analysing networks with NIR may also be valuable when optimising microbes used in bioremediation and bioconversion schemes. Bacteria are extraordinarily flexible respirers, possessing multiple and overlapping pathways for obtaining energy by transferring electrons from high-potential compounds, or electron donors, to lower-potential entities that serve as electron acceptors, such as oxygen and metal ions. For example, Shewanella oneidensis can reduce solubilised heavy metals, such as uranium(VI), to an insoluble form to decontaminate ground water at waste sites, according to Derek Lovley and colleagues at the University of Massachusetts, Amherst.

While S. oneidensis is a remarkably capable organism, the conditions under which it is studied in laboratory cultures differ greatly from those at contaminated sites where it might be used. Indeed, when oxygen is present, S. oneidensis will not reduce uranium(VI). NIR can be used to develop a model of the genetic networks regulating electron transport pathways in S. oneidensis or other microorganisms. In combination with models of bacterial metabolism, such a model could help to manipulate and optimise its performance under field conditions. Similarly, metabolic pathways in other microbes could be manipulated with the help of NIR to enhance their production of valuable compounds, including pharmaceuticals and fuels.

One practical advantage in using NIR is its scalability. Computationally, the NIR algorithm is easily applied to large networks. Experimentally, the scalability of the method depends primarily on the speed with which perturbations can be delivered. So far, we have used only transcriptional overexpressions delivered from episomal expression plasmids, though alternatives such as knockdown approaches based on antisense RNA could also be used. Such perturbations easily could be applied to any gene and require no labour-intensive, biologically unpredictable chromosomal modifications.

Mathematical Models in Microbial Systems Biology

The major challenge in understanding cellular functions from the genomic sequences consists in elucidating and explaining the principles and mechanisms that govern the structures and behaviours of the observed systems. Clearly, data integration is important to enhance the reliability of high-throughput experimentation. System-level data, however, have to be transformed into system-level understanding, which so far is limited. Systems biology addresses this task by combining experimental and theoretical approaches, assigning a central role to mathematical modelling.

Mathematics already enters biology at many levels, from genome sequencing to bioinformatics, where a main objective is to extract statistically relevant patterns. As "no more, but no less, than a way of thinking clearly" about causes and effects, mathematics may, however, make the strongest impact in dealing with, and understanding, complex networks. Mathematical modelling promises to generate experimentally testable hypotheses on underlying mechanisms as well as predictions of cellular behaviour, thereby iteratively producing refined models and insight into the system.

As a general term, it comprises several approaches to representations of reality that are amenable to formal analysis. They range from global, yet coarse, views of cellular systems to very detailed descriptions with a more limited scope. Correspondingly, it is conceivable to follow 'top-down' or 'bottom-up' strategies in experimental analysis and mathematical modelling. Modern experimental technologies enable 'top-down' approaches that start from observations at the network level and successively add detail to initial rough representations.

In the more traditional, yet complementary 'bottom-up' concept, the already available knowledge on biological detail is combined and augmented to establish system-wide models. Recent studies provide clues on the powers and limitations of different types of mathematical models. Here, we can see three major approaches and discuss promising paths to the vision of a complete representation and deepened understanding of microbial cells. The paths have an importance in their own right because of our ability to address fundamental and applied questions in increasingly complex cellular subsystems by combining experimentation and theory.

Function of Network Topologies

Topology—the pattern of interactions between the components—is the most basic feature of any network. In particular, it does not depend on interaction strengths in terms of reaction stoichiometries or velocities (Figure 2a). For instance, genome-scale network topologies were determined from the ensembles of metabolic reactions, of protein-protein interactions, and of genetic dependencies. Graph theory provides quantitative methods for analysing these static representations of components (nodes) and interactions (links).

Interaction-based approaches highlighted the existence of modules, that is, of semi-autonomous units performing distinct functions in cellular systems. Modularity could greatly facilitate a system-level understanding because one can study sub-systems in isolation, and investigate their interactions without having to include all detail of lower levels. A topological analysis of metabolism in 43 organisms suggested metabolic networks to be composed of a hierarchy of modules. Likewise, a large number of studies focused on single levels of cellular regulation. Recent analysis of highly integrated datasets on cellular networks provided more global evidence for hierarchical modularity.

At a detailed level, Alon and co-workers proposed simple, statistically overrepresented interaction patterns in the transcriptional network of E. coli (motifs) as its elementary functional units. Subsequent investigations into transcriptional regulation and/or protein-protein interactions in yeast corroborated this view.

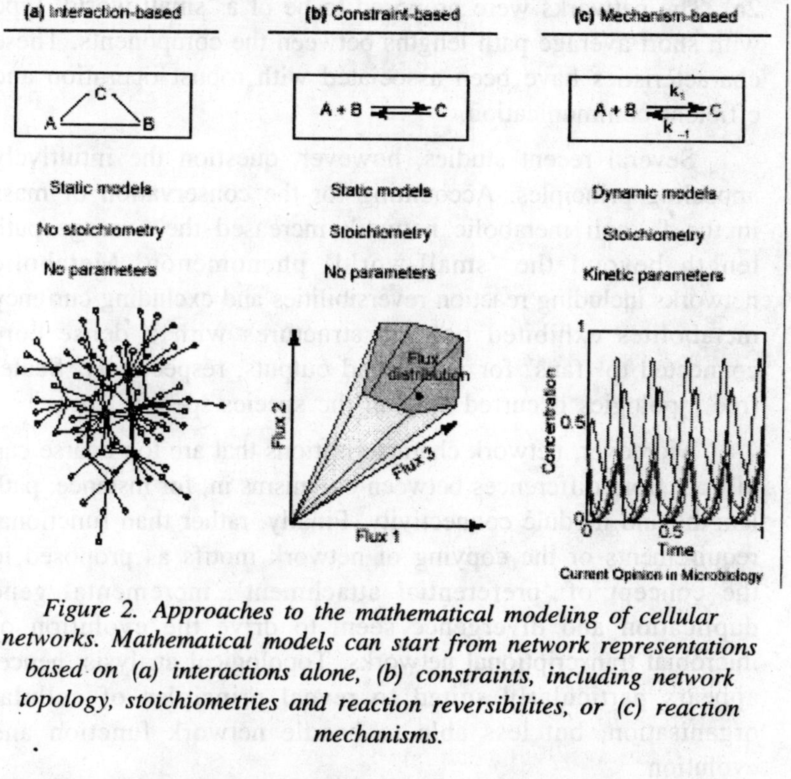

Figure 2. Approaches to the mathematical modeling of cellular networks. Mathematical models can start from network representations based on (a) interactions alone, (b) constraints, including network topology, stoichiometries and reaction reversibilites, or (c) reaction mechanisms.

The potential of motifs, for instance, specifically to speed up transcriptional responses, and evolutionary conservation in network motifs provided independent evidence for a functional role. Moreover, aggregation of motifs in higher-order structures concurs with the big picture of hierarchical modularity as revealed by 'top-down' studies. Exact definitions and unambiguous assignments of modules, however, are largely unresolved issues.

Topological analysis also tries to exploit the fact that the architecture constricts the function and reflects the evolution of networks. Graph theory suggested a 'scale-free' global topology of cellular networks that parallels the structure of complex engineered systems: few highly connected hubs hold the network together, while most nodes have only few connections (Figure 2a). The networks were proposed to be of a 'small-world' type with short average path lengths between the components. These characteristics have been associated with robust operation and efficient communication.

Several recent studies, however, question the intuitively appealing principles. Accounting for the conservation of mass in the E. coli metabolic network increased the average path length beyond the 'small-world' phenomenon. Metabolic networks including reaction reversibilities and excluding currency metabolites exhibited bow-tie structures with a dense core connected to 'fans' for inputs and outputs, respectively. Scale-free topologies occurred only in the species-specific cores.

Moreover, network characterisations that are too coarse can distract from differences between organisms in, for instance, path lengths and module connectivity. Finally, rather than functional requirements or the copying of network motifs as proposed in the concept of 'preferential attachment', incremental gene duplication and divergence seem to drive the evolution of microbial transcriptional networks. Topological analysis, hence, appears particularly suited to reveal principles of cellular organisation, but less able to handle network function and evolution.

Physicochemical invariants such as reaction stoichiometries and reversibilities impose constraints on network function in addition to network topology. In contrast to individual reaction rates, these characteristics are largely available for metabolic networks. Constraint-based modelling exploits them to confine the cellular phenotype to a set of feasible states, namely to all steady-state flux distributions that lie within a convex space of possible solutions (Figure 2b). The rays of the flux cone correspond to minimal functional units ofmetabolism (elementary flux modes or extreme pathways). Their combination characterises all possible behaviours.

With additional hypotheses on optimal network performance, flux balance analysis yields flux distributions for specific conditions. The relatively low data requirements enabled the development of genome-scale metabolic models for several organisms including E. coli and Saccharomyces cerevisiae. Constraint-based analysis of the reconstructed networks leads to experimentally testable predictions such as the (in)viability of knockout mutants. A recent study of yeast metabolism illustrates the scope of the approach. It proved instrumental in predicting the growth phenotype of mutants in approximately 80% of the cases, and (with a lower confidence) qualitative changes in gene expression during diauxic shift.

As the constraints delimit the range of network capabilities, a particular virtue of these models consists in safely identifying the impossible. For instance, incorrect predictions of lethal gene knockouts (unlike 'false positives') reveal inconsistencies in the network structure that can guide model refinement as well as gene annotation for the missing links. Constraint-based analysis requires some sort of network optimality to make specific quantitative predictions. Although the objectives of a cell are generally unknown, the assumption of optimal microbial growth holds in most cases.

For exceptions, combined experimental and in silico analysis revealed that E. coli adaptively evolves to attain the predicted growth capacity. However, it is easy to incorporate alternative hypotheses on network optimality into the constraint-based approach to identify cellular objectives. This type of analysis suggested that E. coli copes with the deletion of metabolic genes by finding a trade-off between improving growth performance and maintaining wild-type flux patterns. Pathway analysis pointed to a control hierarchy that uses gene regulation to establish a general framework for achieving flexibility (which corresponds to robustness) and efficiency, and subsequent enzyme-level fine-tuning for optimal growth.

A corresponding study of yeast central metabolism supported this scheme. Besides inference of optimal control principles frommetabolic network structures, methods for dynamic flux balance analysis, and the use of logical representations of transcriptional regulation as additional constraints extended the scope of the approach. A combined metabolic and regulatory model for E. coli, for instance, proved useful in the iterative knowledge generation on the integrated system by combining model analysis with highthroughput experiments.

Moreover, first concepts for pathway analysis in signaling and gene networks have emerged. Collectively, this promises more generally applicable constraint-based methods that—although limited by predominantly steady-state and non-unique predictions-can facilitate mechanistic modelling efforts.

Mathematical modelling guided by the detailed mechanisms operating in metabolism, signal processing and gene regulation, in principle, leads to detailed, quantitative predictions on cellular dynamics (Figure 2c). Such dynamic models, thus, can help to formulate precise, experimentally testable hypotheses. The complexities of biological systems as well as lacking knowledge on mechanisms and associated parameters have so far limited their success in biology. Moreover, approaching reality by simply increasing the complexity of dynamic models entails extensive computer simulations and rather precludes insights.

To understand complex genotype-phenotype relations, nevertheless, large-scale models founded on mechanistic detail will be indispensable. Recent examples from prototypical microbial systems underline this necessity and provide guidelines for achieving the goal. The lac operon in E. coli as a classic genetic system shows interesting features such as switching behaviour and 'inheritance' of the cell's induction state.

Model-based analysis demonstrated that only the consideration of systemic properties from the molecular to the population level could reproduce and explain the experimentally observed phenomena. This model incorporated the genetic control mechanisms as a module approximated from experimental data. Descending further in scale, an experimental and modelling study of the genetic network showed that multistability occurs only in a certain range of parameter values: a lower binding affinity of the Lac repressor elicits a graded response.

Engineered gene circuits in E. coli provided further evidence for the need of quantitative analysis: networks with identical topology, but different interaction strengths could display qualitatively different signal processing capabilities. Other factors such as intertwined control circuits or cellular heterogeneity could be important in the general case. Modularity, thus, may enable quantitative model building from the genomic to the cellular level, but this crucially depends on module identification.

Bacterial chemotaxis serves as a popular example system for mechanistic modelling since Barkai and Leibler demonstrated that robust adaptation to an unchanged environment results from the network's wiring diagram. A recent comparative study of chemotaxis in E. coli and Bacillus subtilis revealed that homologous proteins form regulatory networks with different interaction patterns in the two organisms, yet establish the same core control strategy. This logic of 'integral feedback control' is necessary and sufficient for robust adaptation. Moreover, additional feedback loops in B. subtilis increase the system's robustness to perturbations.

Comparative model-based analysis, hence, can elucidate principles of cellular regulation resulting from evolutionary processes constrained by the need for robust performance. Mechanistic mathematical modelling, in addition, makes it possible to 'play' with hypotheses on cellular functional or design principles in the computer, and to investigate the effects of alternatives. An intriguing theoretical study dealt with the question of how temporal control of gene expression in metabolic pathways should be designed to achieve fast responses to changing environmental conditions, given a limited amount of enzyme synthesis.

Later experimental and theoretical analysis demonstrated that the predicted type of wave-like expression programme is widely employed by biosynthetic pathways of E. coli, thereby establishing optimal production pipelines for biomass components. At least certain design principles of microbial cells, hence, seem to be accessible through appropriate assumptions on objectives of evolutionary optimisation, in particular, regarding the physiological function and the cellular context of a subsystem. Common themes and integration: modularity, optimality and robustness Roughly, systems biology envisions an understanding of cellular function and behaviour from interactions between genome and environment.

In the domain of mathematical modelling, this translates to formal representations that cover the entire network complexity while maintaining a high level of detail and accuracy (Figure 3). Clearly, none of the approaches discussed herein (and intermediate methods not mentioned) currently has this capability. Mechanism-based modelling is the most obvious candidate for achieving a system-wide understanding, yet not by simple scaling to the genome level.

However, one could aim for a strategy that combines the strengths of present modelling methods with the biologically important themes and highly interconnected issues of modularity, optimality and robustness. In such a scheme (Figure 3), the ultimate outcomes are mechanistic models that, through hierarchical composition of modules and possible abstractions at every level, allow a consistent link between the molecular and cellular level without the need to represent every detail. In principle, this corresponds to ideas on applying engineering principles to understand cells.

In contrast to a pure 'top-down' approach, it would involve iterations between the levels of abstraction. For instance, interaction-based methods could propose modules that, in terms of functionality, would have to undergo tests by detailed quantitative analysis.

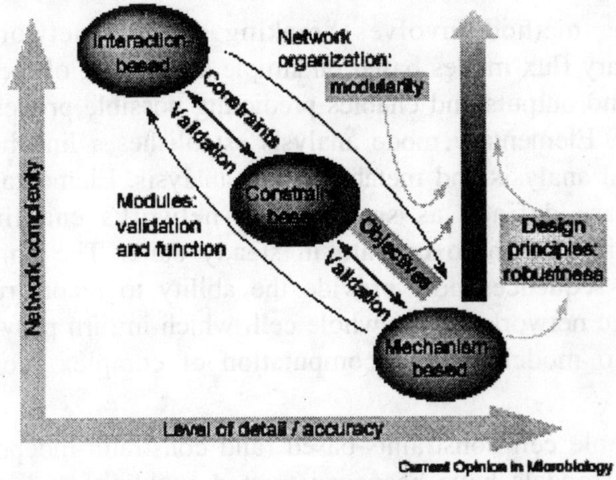

Figure 3. Mathematical modeling: scope and interactions.

Likewise, interactions between the approaches (and comparative analysis of different organisms) could lead to refined concepts of optimality and robustness. General principles, in turn, promise to enhance the prediction capabilities of more abstract models, as documented by the success of cybernetic models. In all these endeavors, a close coupling of theoretical work and experimental analysis is of paramount importance.

Future of Network-structure Knowledge

Prediction Ability Network Structure

The network structures allow predictions using efflux measurements alone that could be experimentally verified for simple organisms. All necessary mechanistic details including kinetic parameters are unlikely to be available for all constituent members of large-scale metabolic networks, the ones that are generally present in the cells. The lack of such mechanistic details renders the dynamic modelling of networks extremely difficult if not impossible. The structure-oriented analyses requiring only network topology have been recently attempted.

The method involves breaking down a network into elementary flux modes based on simple accounting of metabolic inputs and outputs and enables predicting possible properties of network. Elementary mode analysis establishes a link between structural analysis and metabolic flux analysis. Elementary flux modes are defined as smallest sub-networks enabling the metabolic system to operate in steady state. The annotated genome sequences now provide the ability to reconstruct the metabolic network for the whole cell, which in turn provide the ability to modelling and computation of complex biological functions.

Whole cell constraints-based (and constraint-independent) in silico models have been constructed enabling prediction of optimal growth rates on common carbon substrates as well as adaptive evolution that are consistent with experimental data. The knowledge of metabolic pathways have also led to development of number of computational tools for different purposes, network construction as well as prediction. As pointed out that network structure provide the ability to predict the outcomes with minimum available kinetic details pertaining to constituent catalytic proteins. With the knowledge of network structure the mutants becomes a special case of such prediction ability. Mutants also enables test of the ability of predictions. Predicting whether a mutant would be able to grow involves identifying elementary flux modes that do not use the missing reaction.

Non-homologues Metabolic Pathway in Organisms

In many microorganisms the classical metabolic pathway may be different. Understanding of the network structure and comparison of different modules may help predict the intermediate connections in the metabolic pathway. The proteomic methods may help in identifying the interacting partners and define an interactome or module, which would fit with the rest of the pathway components in the organism. For example, in a number of thermophiles, the classical Calvin-Benson pathway members for fixation of CO_2 are absent yet Rubisco is found in a number of members. The proteomic approach may help identify the proteins interacting with Rubisco and thus help determine as to how 3-phosphoglycerate is utilised or how D-ribulose-1,5-bisphosphate is generated in these thermophiles in absence of other members of the classical cycle.

Production of Biochemicals

Currently in chemical as well as in biochemical engineering the approach to produce chemicals are using pure components (reagents or substrates and catalysts). However, with the advent of knowledge of protein complexes, interactomes, modules and network it is increasing becoming clear that future bioengineering would depend on use of interactomes or truncated networks. It is evident that future bioengineering system would involve combination of interactomes or truncated modules of network based on systems design principles derived from the knowledge of system level understanding in biology.

Metabolic engineering embraces the art of joining the different modules or truncated interactomes using the underlying concept of applied system biology, which has been used in pathway engineering to produce new compounds and cell engineering for their use as catalysts in multi-step biotransformations. Although such approaches have been utilised in metabolic engineering in past, however, with the emerging knowledge of global network properties, utilisation of the embedded protocols connecting the interactomes will provide new dimensions to metabolic engineering.

In contrast to chemical and biochemical systems, cells have several properties. A metabolite is often at a branch point and in different organisms these branch points are different. Different proteins act upon the same metabolite in different organisms. It is likely that truncated interactomes from different organisms could be used to generate a novel metabolite from a given metabolite. This will lead to a new path by stabilisation of different but truncated interactomes from different organisms channeling the intermediates in a new direction.

Preparation of Non-living Miniature Multi-step Catalytic Bags

The knowledge of metabolic network or modules and how to keep a portion of that intact, cocooned within the cell may render preparation of non-live cells that may not only allow efficient conversion but may enable extended functioning. Why go to bother of cost intensive purification and immobilisation of enzyme, if the same cohort of enzyme trapped within non-live cells can provide for the same level of conversion. More importantly it is possible that some salvage pathway is retained within the non-live cells enabling efficient conversion using redundant path even if the catalytic ability of one pathway component in the main pathway is impaired.

Although an overwhelming number of microbes are useful for the mankind but a number of them are very harmful as well. They are the pathogens for human, livestock and plants. The fight against pathogenic microbes uses the arsenal of compounds that impairs their metabolic system. However, the tiny microorganisms are extremely versatile, something their multi-cellular counterparts are always lacking. The microbes undergo a selection process against the chemicals used for their destruction. They become resistant to chemicals used against them. The increased resistance to microbicidal agents had been a problem of great concern in medical and pharmaceutical community.

The knowledge of network also confers the ability to attack the system in a much better and concerted manner. For harmful microbes disruption of their metabolism, based on their differences from hosts would be easier and effect with the advent of knowledge of their metabolic network structure and their constituent components. Disruption of metabolic network has been proven to be a successful therapeutic strategy for Mycoplasma, Candida and Leishmania and against plant pathogen Ustilago. A number of recent advances have been made in quantitative experimental approaches, includes new advances in mass spectrometry, electrophoresis with different fluroesent labels.

These proteomic approaches, however, does allow only a static glimpse of abundant proteins. Whereas these powerful approaches has difficulty in identification of proteins at a given point enabling in static glimpse, inside the cell the networks not only identifies the right protein at right time but also enables performing functions so as to enable steady flow of metabolites. How such recognition and functionality is achieved amidst myriads of ensemble of proteins and other components?

The current powerful methods are much away from that understanding. What is needed is equivalent of an ultrasonogram in embryology/neonatology at molecular level. What is needed is a dynamic picture with molecular details where each molecule could be identified separately but at the same time sufficient molecular details of interactions could also be sufficiently decipherable. Quantitative flux analyses approaches to understand metabolome or fluxome have begun, which provides dynamic data and has the potential to provide dynamic information of interactome and modules and eventually that of the network.

In future with superior organised databases are expected to give ability to incorporate information derived from the flux analyses and the kinetic details of the protein components in elemental flux analyses models. Also, based on the static information for the same system analysed under several different conditions may help construction of dynamic maps of the interaction networks. Each proteome analysed under a given condition will be equivalent to a static picture but their combination will enable simulation of dynamic situations. It is now gaining acceptance that understanding secrets of cell lies in understanding complex networks which a cell employs.

The future designs should be putting artificial metabolic pathways and not linear addition of catalysts in bioreactors with all advantage of stability and channeling that is available inside the cells. Such approaches may deliver productivity and rapidity akin to cells and remain impervious to disturbances in operation variables. Stabilisation of such truncated modules (or combining different interactomes) and their being in a state of continuous functionality are important issues.

The knowledge gained from studies on protein interaction and metabolic networks is expected to result in system-level understanding in biology,

— system structure which comprises of gene interaction and biochemical pathways and the mechanisms by which such interactions modulate the physical properties of intra- and inter-cellular structures,

— system dynamics that is how the system responds to external stimuli or operation variables,

— the control methods, the mechanisms that systematically control the state of cell and last and most important

— the design principles or methods.

The design methods are the underlying foundations based on which biological systems have derived their desired properties. The knowledge of cellular design methods would enable bioengineers to simulate rather than blind trial-and-error. The protocols and modules are complementary, interwined and as systems become more complex protocols facilitate the layering of additional protocols involving feedback and signaling. Pathway engineering in bioreactors using underlying design principles would need deeper understanding of system level biology and its implicit layered protocols.

System level understanding of network not only makes better ability to produce new entities or devising new routes for production but also enables novel disruption strategies for treatment of pathogen perhaps using the hitherto known chemical entities. There is a realisation that biological systems has large number of functionally diverse multifunctional interacting set of elements that interact together in a nonlinear fashion to produce coherent behaviour. This has led to approach understanding biological systems as systems, targeting their identification of their structure and unraveling their dynamics. As opposed to controlled complex systems where systems designer often provides target values for the system, in biological systems such targets are created by the system and renewed continuously by the system. Thus there is self-determined evolution in biological systems.

13

PROTEIN INTERACTION NETWORKS

The majority of large-scale interaction experiments have so far been performed on yeast proteins, the past year marks the arrival of the first large-scale animal protein interaction data sets, providing insight into the molecular functions behind multicellularity and cell-cell communication. A large-scale map of approximately 4000 genetic interactions was also derived from synthetic lethal mutations. These interaction data were complemented by several types of supporting data, including large-scale yeast protein localisation data (using GFP-tagged yeast proteins) and the quantitation of the expression levels of approximately 4500 affinitytagged yeast proteins through western blot analysis.

Along with the large-scale data, strategies were refined for more accurate protein interaction mapping, such as the use of isotope labeling techniques to estimate protein enrichment during affinity purification of complexes. Also, many smaller regions of protein interaction networks were explored in detail, such as the EGF (epidermal growth factor) signaling pathway and interactions among bZIP proteins.

Computational methods for predicting interactions also advanced in the past year, with completely new approaches and sophisticated 'mining' of existing interaction data to infer additional interactions. One new trend to be exploited was the tendency of proteins that can functionally substitute for one another to have anticorrelated distribution patterns across organisms, allowing both discovery of non-obvious components of pathways and precise function prediction of uncharacterised proteins.

Another new trend was the tendency of interacting proteins to exhibit similar phylogenetic trees; quantitative algorithms for assigning interaction partners involved analysing trees of families of interacting proteins, such as a ligand and receptor tree, and finding proteins that occupy similar positions in two trees. Computational approaches for predicting novel interactions from known interactions have been developed too: interactions were inferred between pairs of proteins whose sequences are compatible with known X-ray crystal structures of heterodimers and between pairs of proteins with domains that are often observed in interacting proteins.

The specificity of the interactions predicted by the latter approach can, in some cases, be improved by looking for correlated mutations in the domains using an approach dubbed 'in silico two hybrid'. Structure-based interaction prediction (including protein complex prediction and the prediction of crosstalk between complexes) has recently culminated in the delineation of the first network of modeled protein complexes in yeast.

The ability to better predict protein interactions has matured to the point at which online services are now conveniently publicly accessible, such as STRING, PLEX, Bioverse and Predictome. An advantage of such tools is that they allow the prediction of interactions in organisms with no experimental interaction data, permitting systematic searches for new protein systems and the genome-wide characterisation of functional modules.

Although experimentalists appreciate that all data are error prone, strategies for rigorously evaluating the reliability of large-scale protein interaction data sets have emerged only recently. The essential problem is that only a relatively small fraction of interactions in networks are known with any certainty, which leads to difficulties in estimating the rate of both false positives and false negatives. Furthermore, the number of true interactions is considerably larger than the results of typical experiments suggest, implying that the failure of individual experiments to agree on interactions may stem from either poor specificity or poor coverage. In practice, the various interaction data were tested for accuracy on confident sets of interactions.

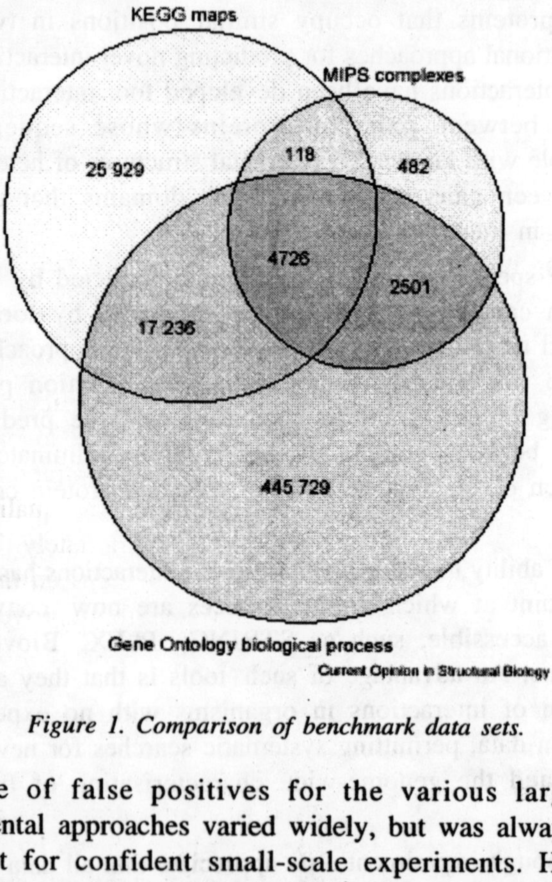

Figure 1. Comparison of benchmark data sets.

The rate of false positives for the various large-scale experimental approaches varied widely, but was always larger than that for confident small-scale experiments. However, highquality subsets could often be chosen on the basis of additional criteria, such as the degree to which mRNAs of interacting proteins are co-expressed in microarray experiments, topological properties of the resulting networks, shared pathways or subcellular localisation, or combinations of these various approaches.

Benchmarks allowing the accuracy of interaction data sets—or, better, individual interactions—to be judged are a prerequisite for the successful integration of data from multiple sources. However, if data of several types are to be integrated, the choice of benchmark set becomes less obvious as not all data will be directly related to physical interactions. This is especially problematic because the agreement among different benchmark sets is surprisingly poor (Figure 1).

Several approaches to integration have been tried, ranging from simple intersections or unions of sets of interactions to more sophisticated probabilistic approaches. The past year also saw the appearance of 'meta-analyses', in which the combination of existing interaction networks suggests additional interactions from the context of the protein network. Two key lessons emerge from the benchmarking and integration studies. First, the measurement of accuracy is critical for the integration of large-scale experiments, which rarely reach the accuracies of experiments done on a small scale. Second, the reconstruction of protein interaction networks seems to be a continuous process in which all data, even noisy data, increase the quality of the network—provided they are integrated appropriately. Thus, the current interaction maps represent the first steps on the way to accurate networks, and should continue to improve in both accuracy and sensitivity.

NEW INTERPRETATIONS OF NETWORKS

The topological properties of protein interaction networks have been intensely studied since the first largescale data sets were published. Interaction networks have been shown by numerous groups to be so-called 'smallworld' networks, an architecture previously observed for several other types of networks. Another topological term frequently attributed to interaction networks is 'scale free'. Although small-world networks need not be scale free, or vice versa, protein interaction networks have both properties.

However, the biological consequences of these topological properties are not clear—in fact, there might not be any, as both small-world and scale-free behaviour can be explained by well-known evolutionary events without the need for any selective pressure acting on the network topology itself. Still, the overrepresentation of genetic interactions between hubs in protein interaction networks supports the hypothesis that hubs play an important role.

In addition to the study of the global topology of interaction networks, the existence of recurring local topological features, known as network motifs, has been shown first in transcriptional networks and later in protein networks. Although some of these motifs make biological sense, the biological significance of network motifs remains to be studied. As interaction networks become increasingly large and complex, there is a growing need to break them down into more manageable subnetworks or 'modules'.

These modules should preferably represent groups of proteins that together contribute to the same cellular function and the modularity should be dictated largely by the topology of the network itself. Functional modules are useful for annotating uncharacterised proteins, for studying the evolution of interacting systems and for getting a general overview of the immediate, first-order functional partners of a protein. Modules are being sought for a variety of networks: metabolic networks, high-throughput experimental interaction data and in silico predicted networks.

Prediction accuracy for the latter type of functional module can be high—almost 90% when benchmarked against manually curated metabolic pathways in Escherichia coli. These modules are thus a rich source for function prediction and pathway annotation. Furthermore, functional genomics data such as mRNA expression profiles can be integrated with high-throughput interaction data to find consistent subnetworks. To identify the actual modules, a variety of supervised or unsupervised clustering techniques are used—often with previous knowledge as the benchmark—but it remains to be seen to what extent the various results are consistent.

In general, functional modules often encompass protein complexes, but conceptually they go beyond stable physical interactions; modules can include transient binding partners and upstream transcriptional regulators, and even proteins that never bind each other but nevertheless function in the same pathway. Implicitly, modules are also the basis of approaches that functionally classify proteins according to their network neighbors. Figure 2 shows two examples of functional modules that expand known protein complexes with either additional subunits or functionally associated proteins. Taken together, identification of network modules not only hints at new cellular systems, but might also guide the ongoing discussions concerning the definition of pathways and cellular processes.

The abundance of interaction data on yeast, and now fly and worm as well, combined with the paucity of information on other organisms, has led naturally to the question of how networks compare between species and to what extent interactions in one organism are maintained in another. Not surprisingly, conserved proteins tend to have conserved interactions and comparison of the yeast network with the Helicobacter pylori bacterial network identified small conserved subnetworks by searching for conserved interactions between pairs of yeast/bacterial orthologs.

A related approach was used to predict a protein interaction map for E. coli from H. pylori interaction data and to import yeast interactions to expand the C. elegans interaction data set, and has led to reconstructions of genetic networks based on the evolutionary conservation of gene co-expression patterns. The significant conservation of interactions con-firms that a feasible strategy for reconstructing networks is to transfer interactions from organisms in which they have been measured.

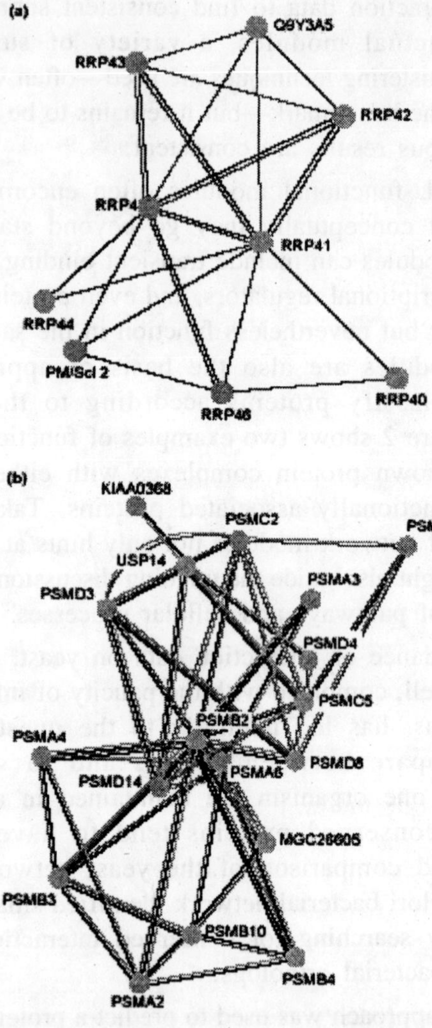

Figure 2. Examples of human functional modules. (a) All proteins but one (Q9Y3A5) are known members of the exosome complex. (b) Proteins in this module form part of the proteasome core and regulatory particles, two of the better characterized protein complexes, as reconstructed from a combination of small-scale and high-throughput protein interaction data in yeast, as well as mRNA co-expression data and phylogenetic profiling using PLEX.

Despite the fact that functional modules are not always present together in distant organisms (i.e. they can change within evolutionary timescales), one can imagine constructing a composite interaction network representing the union of interactions from many different cells within or between organisms, with any particular cell possessing only a subnetwork. However, a realistic characterisation of metazoan interaction networks and their conservation is only now becoming feasiblewith the availability of large-scale data sets, such as the yeast two-hybrid screens in fly and worm.

Human Interactome

Obtaining a reliable interaction set describing the human interactome is a milestone yet to be reached. In contrast to the yeast interactome, the human interactome is largely unknown: a back-of-the-envelope calculation assuming that 3-10 interactions also holds true for each of the 25 000-40 000 human proteins leads to an estimate of roughly 40 000-200 000 interactions. Beyond this obvious uncertainty, this estimate does not even take into account complicating factors such as alternative mRNA splicing or post-transcriptional modification, both of which produce many more protein species and hence more interactions.

Compounding the larger scale of the human interactome is the fact that it has not yet been studied by highthroughput interaction assays and a much smaller fraction of protein interactions are known for human (perhaps 20 000-30 000 total are recorded in the literature) than for yeast. Despite the first medium-scale studies in human centered around individual pathways or machineries, there is a strong need for methods to predict or measure protein interactions that scale to the size of the human interactome. Although experimental approaches are still being scaled to tackle the number of mammalian genes, as witnessed by the first animal protein interaction networks published this past year, computational approaches can rapidly generate initial interaction sets, mostly by transfer of information from other organisms (e.g. Figure 2).

In metazoans such as human, this harbors additional challenges due to the distinct networks in each of the various cell types, many of which have no clear correspondence in other organisms (e.g. there is no adaptive immunity in the fly and probably not too much data should be transferred from the fly exoskeleton to human). As comparative morphology and anatomy is an unfinished research field in its own right, obtaining a thorough and well-annotated benchmark for protein interactions in human is an important next step.

If a common reference was to be accepted by experimentalists and computational biologists alike, from the very beginning of large-scale interaction and network prediction in human, have a much better idea of how much we have to expect. We could avoid misperceptions such as the inflated human gene numbers and could more quickly concentrate on important downstream questions, such as the impact of context on networks, or their temporal and spatial dynamic changes.

14

Implications of Human Metabolome

The human metabolome is a list of all compounds capable of being synthesised by humans. There is no expectation that any human tissues, or even individuals will express their total biochemical potential under any given circumstance, yet there is an expectation that there are a finite number of compounds that can be produced by humans, and that they can be discovered and documented. The human metabolome will be immensely useful in the development and support of the science of metabolomics, and through this medium should yield new perspectives on human physiology. The discoveries that will derive from understanding the full extent and diversity of the human metabolome are likely to be profound.

In other words, the human metabolome is best understood by analogy to the human genome, i.e. where the human genome is the set of all genes in a human, the human metabolome is the set of all metabolites in a human. Whereas the science of genomics is based upon a genome, the science of metabolomics is grounded in a metabolome. To continue the genome/ metabolome analogy, it should be apparent that any published human genome is a statistical approximation as it is derived from a limited number of individuals, and that, ultimately, any individual has a unique genome.

Similarly, the human metabolome is a statistical approximation of the total human metabolic potential. Furthermore, just as the human genome is differentiable from other genomes, for instance, the Xenopus or Caenothus genomes, there is a human metabolome that defines the human biochemical potential that is different from that of Xenopus or Caenothus. While the vitamins and essential amino acids are the most obvious unique characteristics of the human metabolome they are not, by any means, the only defining features.

By its definition, the metabolome is the comprehensive set of all potential metabolites that can be expressed in any human under all conditions, the high-water mark of human biochemical potential. It is likely that most individuals vary in their biochemical potential, expressing only incomplete subsets of the metabolome, depending on their genetic makeup and health state. Indeed, many metabolic diseases and even the efficacy of most drugs are variable, due, at least in part, to individual variances in metabolism and the resulting biochemistry.

Fundamentally, the concept of a metabolome is based in set theory. The metabolome is the total set, and all individuals, tissues, and conditions are subsets. What is unique about the metabolome is its ability to directly reflect physiological status when used as part of a metabolomic analysis.

METABOLOMICS

Metabolomics, also known as biochemical profiling, is an emerging "omics" science that has tremendous potential and many unique characteristics. It is most obviously useful for the exploration of problems in which physiology is altered, e.g. through stress, disease, chemical or other insult. Like all "omics" sciences (Fig. 1) it takes a global view of an organism, i.e. attempting to understand the current physiological status of a sample or organism in light of its full physiologic potential.

Thus, genomics looks at the total genome to understand the significance of an individual gene. Similarly, transcriptomics and proteomics attempt to understand change as a function of individual transcript expression or protein concentration in light of the total expression of RNA transcripts or proteins, respectively. Despite the strength of these technologies, deriving from their breadth of information, in these three cases the current limitations to interpretation lie in the fact that the vast majority of the relevant entities (i.e. genes, transcripts and/or proteins) are only partially characterised.

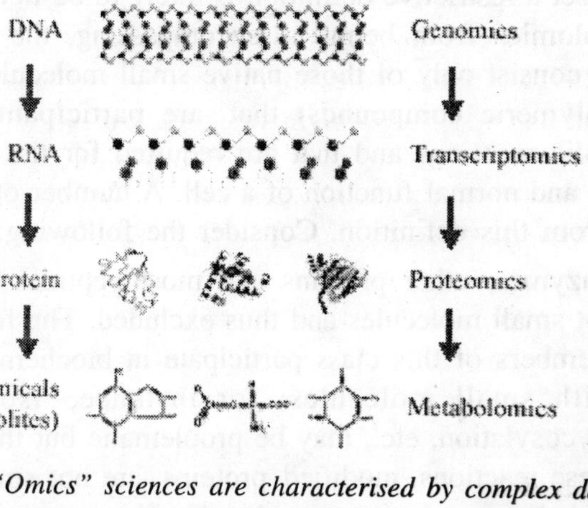

Figure 1. The "Omics" sciences are characterised by complex datasets of related phenomena each of which taken as a whole constitute a picture of an organism.

In other words, while their respective sequences may be known, their functional significance is often largely hypothetical, if understood at all. These sciences, responsible for huge advances in the past few years, are comparative in nature and, although understood through "analytical" technologies, they are not truly quantitative. They usually achieve their successes by comparing one or more states to determine where change has occurred. In this light, the "omics" sciences are generally hypothesis-generating, and form the basis for specific "non-omics" investigations that are supported by classical experiments.

Metabolomics is a parallel science in that it characterises the physiological state of a sample by determining the "concentration" of all of the small molecules that constitute metabolism. In contrast to the other "omics," however, the nature and the relationship of almost all of the metabolomic entities (i.e. biochemicals) have been thoroughly established through over a century's worth of biochemical investigations. The establishment of a metabolome is fundamental to a full understanding of metabolomics.

Human biochemistry is very complex, enough so that at the outset a restrictive definition is likely to be helpful. To keep metabolomics from being all-encompassing, the metabolome should consist only of those native small molecules (definable, non-polymeric compounds) that are participants in general metabolic reactions and that are required for the maintenance, growth and normal function of a cell. A number of implications flow from this definition. Consider the following:

— Enzymes, other proteins and most peptides are generally not small molecules and thus excluded. The fact that many members of this class participate in biochemical reactions with small molecules; for instance, isoprenylation, glycosylation, etc., may be problematic but the products of these reactions, modified proteins, are not small molecules and thus are better considered within proteomics. In their construction and degradation they consume or yield small molecules, and thus, have locatable and specific output or input points to the metabolome.

— Genetic material (all forms of DNA and RNA) is also excluded according to size and function. Furthermore, these are better considered within their own disciplines of genomics and transcriptomics. In its construction and degradation it consumes or yields small molecules and, thus, it also has locatable and specific output or input points to the metabolome.

— Structural molecules (Glycosaminoglycans, and other polymeric units) similarly may be built up from and degraded to small molecules but do not otherwise participate in metabolic reactions.

— Polymeric compounds, such as glycogen, are important participants in metabolic reactions but are not chemically defineable. Thus, like all the previous categories, they represent a source of metabolites, i.e. an input/output to metabolism, but are not metabolites themselves.

— Metabolites of xenobiotics are not native, required for the maintenance, growth or normal function of a cell, and thus are not really part of the metabolome. Their significance, not to be underestimated, is the realm of toxicology.

— Essential or nutritionally required compounds are not synthesised de novo, (i.e. not native), but are required for the maintenance, growth or normal function of a cell; therefore, they are part of the metabolome as inputs from nutrition.

While arguments could be made for the inclusion of these, or the exclusion of other categories, to do so does not serve metabolomics, as it would shift the focus away from metabolism, and, with it, from physiology. At some point a general catalogue of all of the components of a cell, inclusive of genomic, transcriptomic, proteomic, metabolomic and other molecular entities will need to be collated and a metabolome may be considered a first step in this direction. Such a general catalogue would begin to approach the natural ecology of all molecular space. It will undoubtedly be a powerful perspective that will build upon current ongoing efforts, of which the establishment of a metabolome will be but one first step.

Finally, despite efforts to be limiting in the previous paragraphs, biochemicals and metabolites are generally not viewed as congruent terms. As a matter of historical precedence, the term "metabolites" is often mistakenly taken to indicate only that subset of all metabolites that are part of degradation pathways. It is important to reconcile this by recognising the broadest possible definition of the term as the basis of metabolomics and metabolomes.

Construction of Metabolome

One way to deduce the human metabolome is by metabolic reconstruction from the human genome. Such an effort would mine a variety of publicly accessible and proprietary databases to compile a listing of compounds that are within the potential of a human being to biosynthesise. Since the metabolome is developed from the human genome, it will not include xenobiotics or their metabolites.

Metabolic reconstructions have been widely used for the reconstruction of microbial genomes, but not yet on a more complex organism. The reconstruction will first scan the annotations from the Human Genome for all enzymes (Fig. 2). This starting point, a preliminary list of enzymes, will be qualified to ascertain that they are all appropriate. Those enzymes that emerge will be informatically translated to a preliminary list of reactions expected to occur by reference to a metabolic database. This list of reactions defines a starting list of compounds that act as either substrates or products for these "assumed-to-bepresent- in-humans" reactions.

These compounds, as they are newly identified, are inserted into the Human Metabolome Database. At this point a return cycle begins which looks at the preliminary set of metabolites and asks in what other reactions they are known to participate. This examination suggests a list of proposed reactions which yields a list of proposed enzymes. The known sequences for these enzymes are then sought in the Human Genome yielding a continuation cycle. These continuation cycles are reiterated until as much of primary metabolism as possible can be uncovered by genomic probing. When there is sufficient understanding of the framework of metabolism, then a new strategy may be employed, namely the proposal of whole pathways that are suggested by the so-far-discovered collection of reactions. This yields additional proposed enzymes that become a source of additional continuation cycles.

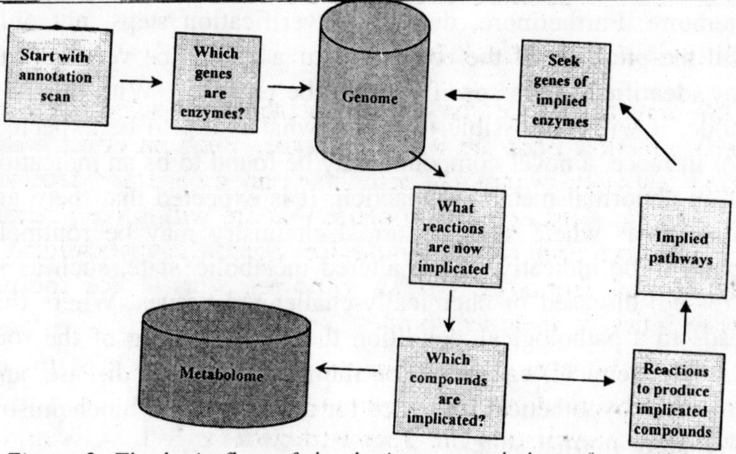

Figure 2. The logic flow of developing a metabolome from a genome.

First determine all known enzymes in a genome. This implies a preliminary set of precursors and products. These each imply additional enzymes which are then sought in the genome. As enzymes are identified their associated chemistry may be assumed to be part of the metabolome of that organism. In this way it is expected that the majority of human biosynthetic capacity may be understood from first principles; certainly, all of the core metabolic intermediates should be identified. Beyond this point additional methods may be needed to fill in more peripheral pathways. These will likely be added by literature-based curation, a more strenuous and often less reliable method.

A benefit of this reaction-based approach will be that all of the metabolites in the metabolome will be associated with one or more enzymes, and will fit into known biosynthetic relationships. Other approaches have been proposed which are not reaction-based, but rather chemistry based. These suffer from the drawback of being just lists of disjointed compounds. The completion of such a project will be the development in one place of a scientifically validated Human Metabolome, i.e. a list of compounds to be expected in a human, irrespective of location.

Clearly every compound that is referenced in the Human Metabolome will need to be verified in the scientific primary literature. Furthermore, during the verification steps, not only will the presence of the compound in a human be verified, but any identified tissue specificity will be recorded. With this as a guide, it will be possible to define what is not to be expected. For instance, a novel compound may be found to be an indication of an abnormal metabolic reaction. It is expected that there are some areas where such abnormal chemistry may be routinely found to be indicative of an altered metabolic state, such as in stressed, diseased or chemically-challenged tissues. Where this leads to a pathological condition the understanding of the root (i.e. biochemical) cause will be illuminating to the disease, and provide a hypothetical reference for a return of the biochemistry to a more normal state.

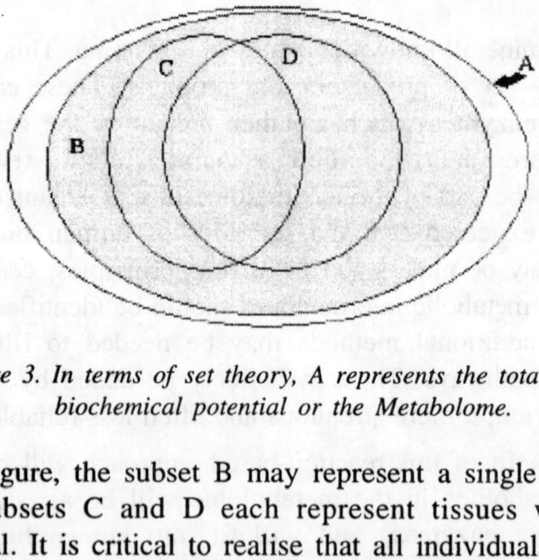

Figure 3. In terms of set theory, A represents the total human biochemical potential or the Metabolome.

in this figure, the subset B may represent a single individual, while subsets C and D each represent tissues within that individual. It is critical to realise that all individuals are likely to be represented by different subsets and that all tissues are in a constant state of flux and thus C and D likely vary, to some minimal extent, over time.

From experiments that have already been done in other species it is clear that, just like the relationship between the transcriptome and a gene array pattern, the number of all compounds likely to be expressed in any human tissue will be a subset of the full metabolome. Thus, the biochemical profile of a liver is different from the biochemical profile of a muscle (Fig. 3). Even though the biochemical profile of a tissue is generally stable and reproducible, and represents the baseline metabolic status of that tissue over time, there is some absolute drift in physiological status. By extension of this thought, when a tissue is diseased, the biochemical status of the diseased tissue must reflect the altered physiological status. The usefulness of the complete human metabolome in understanding these issues will be immense.

Preliminary estimates of the total number of compounds in a human have been varied. The standard wallchart of metabolism, which lists reactions that are not present in humans, lists only about 800 compounds in core primary metabolism. Most biochemical textbooks extend this list to no more that 1200 to 1500 compounds, again drawing from all life forms. By extensive querying of publicly available databases we have been able to extend this list, limited now to humans, to no more than 2000 compounds. This number of compounds will form a very workable and firm foundation for the evolving science of metabolomics.

15

METABOLIC NETWORK PREDICTION

Functional genomics is the search for understanding of the functionality of specific genes, their relations to diseases, their associated proteins and their roles in biological processes. In functional genomics, a cell can be seen as a biochemical machine that consumes simple molecules to generate more complex ones by chaining together biochemical reactions into long sequences; its processes are referred to as metabolic pathways. Genes play an essential role in these networks by providing the information to synthesise the enzymes that catalyse biochemical reactions.

Metabolism is a set of chemical reactions, used by living organisms to process chemical compounds in order to take energy, extract building blocks and eliminate toxic compounds. Most of these reactions would not be executed without specialised proteins called enzymes, whose function is to catalyse these chemical reactions. Metabolism was previously seen as a combination of distinct pathways, such as glycolysis, citrate cycle, urea cycle, amino acid biosynthesis and many others. All these pathways are connected to each other.

In recent years, metabolism has started to be studied in a network approach. Information about the structure of metabolic networks can now be partially extracted and represented in graphical form using KEGG, WIT and MetaCyc. For example, the data of the Kyoto Encyclopaedia of gene and Genomes (KEGG) consists of information on interacting molecular and gene pathways. Related to KEGG are the Biochemical Pathways (BP) index of Boehringer Mannheim and the Encyclopaedia of E. Coli Genes and Metabolism (EcoCyc).

Enzymes are proteins encoded by genes and these genes can be expressed at will. Therefore some subgraphs of the network or pathways can be activated or inactivated. In prokaryotic cells, a whole subgraph can be activated or inactivated at once, this is the notion of operons; and in eukaryotic cells which do not have operons, genes can be controlled individually allowing an even thinner regulation. The whole network is very dynamic, responding to the environment and the cell's needs, and some parts of the network can be activated in mutually exclusive or inclusive way. If we look at the level of a single biochemical reaction in the network, its activation depends on the presence of the reaction substrates, therefore depends on the previous step. It also depends on whether the gene coding for the enzyme is activated, and of course what activates the gene, which can be one of the substrates or some external stimuli.

Causal Networks Construction

A Bayesian network is a tool for representing a probability function. It is defined over a finite domain V of variables, each of which may be discrete or continuous. A Bayesian network consists of a DAG G whose nodes are the variables in V; a probability specification S which contains the probability distribution $p(V_i|Par_i)$ of each variable $V_i \in V$ conditional on its parents Par_i in G; and an assumption, called the Markov condition, which states that each variable $Vi \in V$ is probabilistically independent of its non-descendants, ND_i, conditional on its parents, written $V_i - ND_i|Par_i$.

A causally interpreted Bayesian network, or causal network, is a Bayesian network in which the graph G represents the causal relations amongst the variables in V, with an arrow from V_i to V_j if V_i is a direct cause of V_j. There are arguments to the effect that, if a Bayesian network is causally interpreted then the Markov condition—now called the causal Markov condition—is a valid assumption: i.e. that any variable is probabilistically independent of its non-effects, conditional on its direct causes. Thus in cases where causal relations are known, the graphical structure G in a Bayesian network can just be taken to be the causal graph. This reduces the problem of constructing a Bayesian network to that of determining the probability specification S, and this is normally done by taking the corresponding sample frequencies from a database of past case data, or by eliciting degrees of belief from experts.

In the case of yeast *saccharomyces cerevisiae* metabolism the network of aromatic amino acid pathways takes the form of a causal graph (Figure 1). The direction of causation (*i.e.* the direction of the reaction) is from the top to the bottom of the diagramme. The rectangular nodes are enzymes and the circular nodes are metabolites. The white circular nodes are the aromatic amino acids—phenylalanine (C00079), tyrosine (C00082) and tryptophan (C00078). The values that the variables take are their concentrations (by mass). C00079 is produced by C00166 and C00025 under enzyme 2.6.1.7.

There is a period of flux before each reaction settles down to an equilibrium during which the reaction often takes place in both directions. There is inevitably a single overall direction to the reaction however, which is determined after using a conditional dependency measure to test for violations of the Markov condition.

In KEGG, the metabolic pathways represent all known pathways in a given organism. However, when, in some case, enzymes could not be located, a simple deductive rule was used to uncover alternative reaction paths from an initial substrate and a final product. However, Goto describes difficulties in path computation from a given list of enzymes—there still exists a number of unknown pathways for secondary metabolisms and metabolisms that are revealed under stressful conditions.

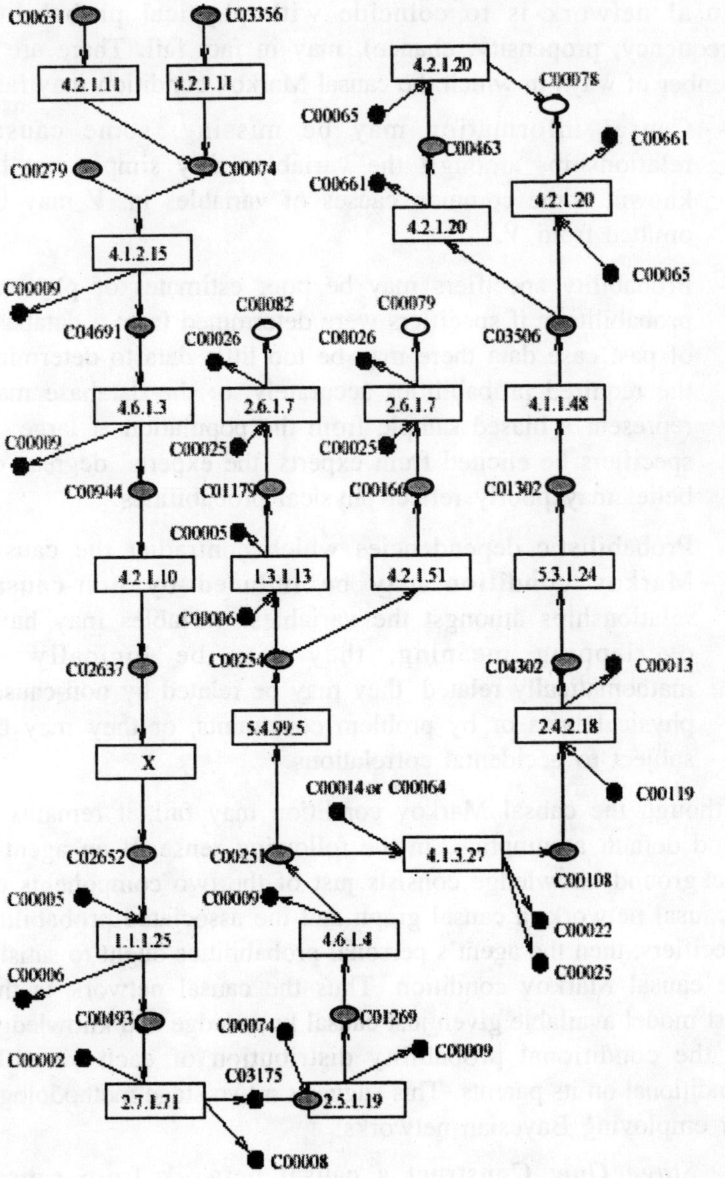

Figure 1. A Causal graph of metabolic pathway of aromatic aminoacid of yeast Saccharomyces cerevisiae

The causal Markov condition, which is required to hold if the causal network is to coincide with physical probability (frequency, propensity, chance), may in fact fail. There are a number of ways in which the causal Markov condition may fail:

— Causal information may be missing: some causal relationships amongst the variables may simply not be known; some common causes of variables in V may be omitted from V.

— Probability specifiers may be poor estimates of physical probabilities: if specifiers were determined from a database of past case data there may be too little data to determine the required probabilities accurately, or the database may represent a biased sample from the population at large; if specifiers be elicited from experts, the experts' degrees of belief may poorly reflect physical probabilities.

— Probabilistic dependencies which contradict the causal Markov condition may be induced by non-causal relationships amongst the variables: variables may have overlapping meaning, they may be logically or mathematically related, they may be related by non-causal physical laws or by problem constraints, or they may be subject to accidental correlations.

Although the causal Markov condition may fail, it remains a good default assumption, in the following sense. If an agent's background knowledge consists just of the two components of a causal network, a causal graph and the associated probability specifiers, then the agent's personal probabilities ought to satisfy the causal Markov condition. Thus the causal network is the best model available given just causal knowledge and knowledge of the conditional probability distribution of each variable conditional on its parents. This suggests a two-stage methodology for employing Bayesian networks:

— *Stage One*: Construct a causal network from causal knowledge and corresponding probability specifiers. This is a good default model.

— *Stage Two*: If the network fails to perform well (this is indicative of failure of the causal Markov condition), modify the network so that it better approximates physical probability.

MODIFICATION OF NETWORK

If a Bayesian network is to be modified to better represent physical probability, one can either change its graphical structure, or its probability specifiers or both. Changing probability specifiers to better approximate the corresponding physical probabilities is a statistical problem. This problem is solvable—i.e. a mechanism is available for determining physical probabilities. Two routes are available for changing the graph in a Bayesian network: one can change the nodes in the graph (add, delete or combine nodes) or change the arrows in the graph (add, delete or re-orient arrows) or both. Here present an example of each strategy: adding arrows and adding hidden nodes.

The adding-arrows approach is conceptually very simple. If one adds an arrow from V_i to V_j in a Bayesian network (and change the corresponding probability specifiers accordingly) then the new network will be no worse an approximation to physical probability than the old network, and will be a closer approximation if and only if V_i and V_j are probabilistically dependent conditional on the other direct causes of V_j. So a simple strategy for changing a network to better approximate physical probability is to add arrows corresponding to conditional dependencies. If at each stage one adds the arrow corresponding to strongest conditional dependence then one achieves the closest approximation at each stage. Moreover this simple greedy algorithm finds networks that are close to the global best approximation.

Whenever two nodes *B* & *C* with no arrow between them are probabilistically dependent conditional on a common parent *A* (a violation of the causal Markov condition), then a 'hidden node' *H* is added as a new parent of *B* and *C*, with the arrows from A to B and C redirected through H. Then probability specifiers for H, B and C will be learned from data using the symmetric propagation algorithm called Symmetric Hidden Node Method (SHNM). In neural cell morphology, SHNM improved the prediction accuracy in Bayesian networks up to 42% (from 59% to 84%). Comparative analysis on other machine learning techniques also showed the strength of SHNM. These included neural networks and C4.5. The neural networks had one hidden layer (with up to 5 hidden nodes).

The number of learning cycles was in the range 10,000 to 500,000, compared to 800 cycles for learning neural networks and Bayesian networks, respectively. The C4.5 weights were set in the range 2 to 4. It showed that the C4.5 method gave a comparable performance to the naive Bayesian network, but neural networks were considerably worse in this case.

According to the principle of the common cause a probabilistic dependency which violates the causal Markov condition indicates that either a direct causal relation between the dependent nodes, or a common cause of the two nodes, is missing from the causal graph. Thus to generate a causal network that satisfies the causal Markov condition, need the flexibility to add either a new arrow or a new common cause (a hidden node). The new graph can be treated as a new causal hypothesis, and can motivate closer scrutiny to verify the new posited causal connections.

In deciding whether to add an arrow or add a hidden node to modify a network, there are two key considerations to take into account. First the new network should be plausible when construed as a hypothesis about causal relations. Thus if it is implausible that two dependent variables are directly causally related, one ought not add an arrow between them—one ought to add a hidden node (interpreted as a common cause) to account for their dependency. Second, one ought to pursue the option that, other things being equal, increases the complexity of the network least. The complexity of a network can be measured in terms of the number of probability specifiers required in the network. In most situations adding an arrow will increase complexity least, but in cases where two or more nodes share a large number of parents, adding a hidden node can even decrease complexity.

Objectives for Yeast Metabolism

Methodology for network modification has two objectives: Firstly, to significantly improve the prediction accuracy of the causal network. Secondly, to suggest new common causes (chemical reactions) and causal relations (reaction pathways between substrates and products) in the network. The algorithm is as follows

— For each pair of variables (substrates and products) *B* and *C* in the network, check their probabilistic dependence conditional on the parents *A* of *C*, via the mutual information formula:

$$MI_c(B,C|A) = \sum_{B,C}\left[P(b,c|a)\log\frac{P(b,c|a)}{P(b|a)P(c|a)}\right]$$

— If there are any such dependencies then the causal Markov condition has failed. Choose the maximal dependency and generate two new causal hypotheses: one by adding a hidden node and the other by adding an arrow from *B* to *C*.

— The corresponding probability specifiers need to be learned. In the case of the model with added arrow these can be equated with the corresponding frequencies in the database. In the case of the model with hidden node, execute a learning process in that particular local structure.

— Try to find the referent of a hidden node *H* as follows. Regenerate a data set for *H* relevant to the original data set. Using a pattern matching techniques with correlation calculation, try to locate a variable with unknown functionality that scores the highest mark.

— Try to verify a new arrow by checking that it corresponds to probability raising of the effect by the cause, conditional on the effects other direct causes. If so, check that intervening to fix the value of the cause fixes the values of the effect, controlling for the effect's other direct causes.

— Eliminate a hypothesis if it has no plausible causal interpretation.

16

ALGEBRAIC COMPARISON OF METABOLIC NETWORKS

Large scale bacterial phylogenies that are based on single genes are notoriously plagued by gene transfer, gene duplication, gene deletion, and functional replacement of genes. For this function, the metabolic networks of a wide variety of organisms, in particular procaryotes, have been reconstructed by means of a combination of genomic annotations with biochemical and physiological data.

Christian V. Forst. et.al conducted a study with 16S RNA phylogenies, to determine the algebraic comparison of metabolic networks, phylogenetic inference, and metabolic innovation. Details of ther study are discussed below.

A metabolic network is defined by its metabolites and the system of reactions that inter-convert them. Researchers denote the set of metabolites by X. A chemical reaction can be described as a pair of multisets (E^-,E^+), where $E^- \subseteq X$ is the set of educts in the reaction and $E^+ \subseteq X$ is the set of reaction products.

Slightly more generally, they can replace the multisets by an ordinary sets and instead define the multiplicities of product and educt metabolites by means of the stoichiometric coefficients n^+_x, E and n^-_x, E of the products and educts, respectively. A metabolic network is thus a pair (X, ε) where ε is a set of reactions. Such a structure is known as a directed hypergraph $\mathfrak{M}(X, \mathcal{E})$,. The stoichiometric matrix S of the network has the entries

$$S_{xE} = n_{x,E} — n^{-}_{x,E} \tag{1}$$

For completeness, they remark that the set $E^c = E+\cap E^-$ are the catalysts of the reaction E. Furthermore, a reaction is autocatalytic if $n^+_{x,E}—n^-_{x,E}\neq 0$ for some $x \in E^c$. By abuse of notation they write $E = E^+ \cup E^-$ for the set of metabolites involved in the reaction E. Furthermore, they write supp $\mathcal{E} = \cup\{E\in E\,\mathcal{E}\}$ for the set metabolites that actually take part in the reactions. They call a network $\mathfrak{M}(X, \mathcal{E})$ clean if $X = \text{supp}\mathcal{E}$ and define the clean up operator as $\lfloor\mathfrak{M}\rfloor = (\text{supp }\mathcal{E}, \mathcal{E})$. Furthermore, for a given set$\mathcal{E}$ of reactions and set A metabolites they define

$$\mathcal{E}[A] = \{E\in \mathcal{E}\backslash (E^+\cup E^-) \subseteq A\} \tag{2}$$

The restriction of a network $\mathfrak{M}(X,\mathcal{E})$ to a set A of metabolites is defined as the clean network

$$\mathfrak{M}[A] = \lfloor(A, \mathcal{E}[A])\rfloor \tag{3}$$

For short they write $\mathfrak{M}[\mathcal{E}] = \mathfrak{M}[\text{supp}\mathcal{E}]$ for the restriction with respect to a set of reactions. The number of reactions in a network$\mathfrak{M}$ will be denoted by $||\mathfrak{M}||$. In order to compare networks in a systematic way, they need to be able to determine the differences and the commonalities of two networks. In the following let $\mathfrak{M}'(X', \mathcal{E}')$ and $\mathfrak{M}''(X'', \mathcal{E}'')$ be two networks. Of course they have $\mathfrak{M}' = \mathfrak{M}''$ iff $X' = X''$ and $\mathcal{E}' = \mathcal{E}''$. The empty network will be denotes by$\varnothing$.

Union: The union $\mathfrak{M} = \mathfrak{M}' \cup \mathfrak{M}''$ is defined as the network $(X' \cup X'', \mathcal{E}' \cap \mathcal{E}'')$. Note that $\mathfrak{M}$ is clean if both $\mathfrak{M}'$ and $\mathfrak{M}''$ are clean.

Intersection: The intersection $\mathfrak{M} = \mathfrak{M}' \cap \mathfrak{M}''$ is defined as the clean network

$$\mathfrak{M} = \lfloor (X' \cap X'', \mathcal{E}' \cap \mathcal{E}'') \rfloor \tag{4}$$

Note that $(\mathcal{E}' \cap \mathcal{E}'')[X' \cap X''] = \mathcal{E}' \cap \mathcal{E}''$.

Difference: The difference $\mathfrak{M} = \mathfrak{M}' \setminus \mathfrak{M}''$ is defined as the clean network

$$\mathfrak{M} = \lfloor (\mathrm{supp}(\mathcal{E}' \setminus \mathcal{E}''), \mathcal{E}' \setminus \mathcal{E}'') \rfloor \tag{5}$$

The difference Network contains all reactions occurring in $\mathfrak{M}'$ but not $\mathfrak{M}''$, and all metabolites occurring in the remaining reactions.

The strict difference $\mathfrak{M} = \mathfrak{M}' \setminus\setminus \mathfrak{M}''$ is the clean network

$$\mathfrak{M} = \lfloor (X' \setminus X'', (\mathcal{E}' \setminus \mathcal{E}'')\, [X' \setminus X'']) \rfloor \tag{6}$$

The new network contains only those metabolites occurring in $\mathfrak{M}'$ but not $\mathfrak{M}''$, and only those reactions from $\mathfrak{M}'$ that can be performed with the remaining metabolites. Thus, they have $|| \mathfrak{M}' \setminus\setminus \mathfrak{M}'' || \leq || \mathfrak{M}' \setminus \mathfrak{M}' ||$.

Symmetric difference: The symmetric difference $\mathfrak{M} = \mathfrak{M}' \Delta \mathfrak{M}''$ is defined as the clean network $\mathfrak{M} = \lfloor (\mathfrak{M}' \cup \mathfrak{M}'') \setminus (\mathfrak{M}' \cap \mathfrak{M}'') \rfloor$.

Strict symmetric difference: The strict symmetric dfference $\mathfrak{M} = \mathfrak{M}' \Diamond \mathfrak{M}''$ is $\mathfrak{M} = \lfloor (\mathfrak{M}' \cup \mathfrak{M}'') \setminus\setminus (\mathfrak{M}' \cap \mathfrak{M}'') \rfloor$.

The Vienna Reaction Network Library Vienna-RNL implements these basic set-theoretic operations on chemical reaction networks. It provides basic ANSI C data structures for chemical reactions and their networks, IO routines for reading and writing and various formats, as well as set operations such as the union, intersection, or difference of two chemical reaction networks. It is intended for the use in conjunction with the user's own Cprogrammes or PERL scripts.

An extension of the IO-Routines for reading and writing SBML, an XML based dialect for the standardised representation of systems biology models, is currently being implemented. The capability of reading and writing SBML will make the functionality of the Vienna Reaction Network Library accessible to about 30 other software systems 2 which support SBML.

Phylogenies from Networks

Datasets were retrieved from the KEGG database on metabolic networks, which holds genomic and network data of about 20 Archaea, 200 Bacteria and 20 Eucarya, where in particular the data of many Eukaryotes are incomplete. In a preparatory step they decomposed the individual KEGGpathways into their chemical reactions and combined these to a complete network for each organism.

The simplest approach to inferring phylogenetic relationships form metabolic networks is to use a distance measure d on the set of reaction networks. They use here

$$\begin{aligned} d(\mathfrak{M}', \mathfrak{M}'') &= \frac{\|\mathfrak{M}' \triangle \mathfrak{M}''\|}{\|\mathfrak{M}'\| + \|\mathfrak{M}''\| - \|\mathfrak{M}' \cap \mathfrak{M}''\|} \\ &= \frac{\|\mathfrak{M}' \triangle \mathfrak{M}''\|}{\|\mathfrak{M}' \cup \mathfrak{M}''\|} \end{aligned} \tag{7}$$

Alternatively, the strong symmetric difference $\mathfrak{M}' \Diamond \mathfrak{M}''$ could be used to de-fine a difference measure. Furthermore, other normalisations of the difference measure could be used. They have observed, however, that equ.(7) performs best with respect to reproducing trusted 16S RNA phylogenies. Distancebased network phylogenies are computed using the Fitch algorithm

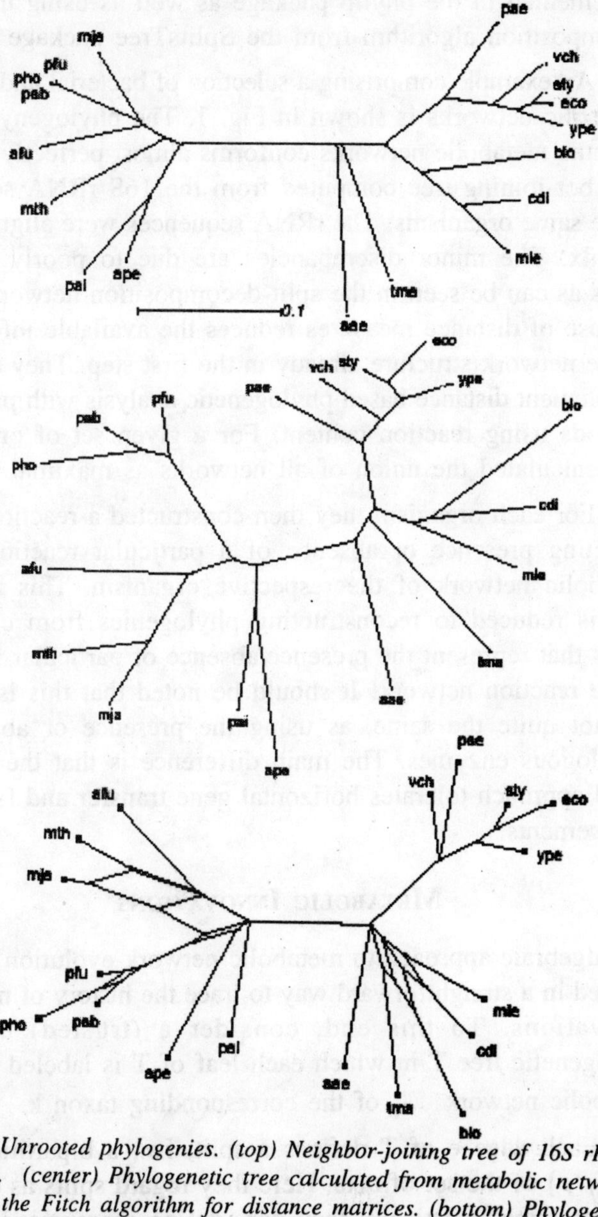

Figure 1. Unrooted phylogenies. (top) Neighbor-joining tree of 16S rRNA sequences. (center) Phylogenetic tree calculated from metabolic network data using the Fitch algorithm for distance matrices. (bottom) Phylogenetic tree calculated from metabolic network data using Splits decomposition with the Fitch-Margoliash power 2 fit for distance matrices.

implemented in the phylip package as well as using the splits-decomposition algorithm from the SplitsTree package.

An example comprising a selection of bacterial and archaeal metabolic networks is shown in Fig. 1. The phylogeny inferred from the metabolic networks conforms almost perfectly with the neighbor-joining tree computed from the 16S rRNA sequences of the same organisms. The rRNA sequences were aligned using clustalx. The minor discrepancies are due to poorly resolved nodes as can be seen in the split-decomposition network below. The use of distance measures reduces the available information on the network structure already in the first step. They therefore complement distance based phylogenetic analysis with parsimony methods using reaction content: For a given set of organisms, they calculated the union of all networks as maximal network.

For each organism they then constructed a reaction profile reflecting presence or absence of a particular reaction in the metabolic network of the respective organism. This approach thus is reduced to reconstructing phylogenies from character-tables that represent the presence/absence of particular reactions in the reaction network. It should be noted that this is similar, but not quite the same, as using the presence or absence of orthologous enzymes. The main difference is that the network based approach tolerates horizontal gene transfer and functional replacements.

Metabolic Innovations

The algebraic approach to metabolic network evolution can also be used in a straightforward way to trace the history of metabolic innovations. To this end, consider a (trusted) unrooted phylogenetic tree T in which each leaf of T is labeled with the metabolic network $\mathfrak{M}_k$ of the corresponding taxon k.

Each edge e of T defines a split, i.e., a bipartition σ_e = {U_e, $\bar{U}$ e} of the set of taxa. Here they regard splits as directed. Note that mathematically they can define innovations at each split in both directions. One of the two subsets U or $\bar{U}$, however, contains the ancestral state, hence only one direction makes biological sense: this is the one where the ancestral state (root of the tree) is located in the sub-set $\bar{U}$. This knowledge has to be provided externally.

Consider an (arbitrary) directed split $\sigma = (U, \bar{U})$ on the given set of taxa, i.e., a pair of sets of taxa $(U, \bar{U})$ such that $U \neq \phi, \bar{U} \neq \phi$, and $U \cap \bar{U} = \phi$ They define the differential metabolic network

$$\mathfrak{D}(\sigma) = \left(\bigcup_{k \in U} \mathfrak{M}_k\right) \setminus \left(\bigcup_{k \in \bar{U}} \mathfrak{M}_k\right) \tag{8}$$

The network $\mathfrak{D}(\sigma)$ describes the metabolic innovations in U relative to the "background" $\bar{U}$

Network phylogenies are rather sensitive with respect to life-style and environmental constraints. The organisms whose metabolic networks have been utilized to compute the 16s rRNA tree shown in Fig. 1 are capable to freely live in the environment with a reasonably large capacity for adaptation. As a first example they analysed the unique metabolic network from the Pyrococcus genus.

Figure 2 shows the network phylogeny from Fig. 1 with the Pyrococcus spp. clade highlighted. The resulting differential network indicates reactions present in Pyrococcus spp. but absent in all other organisms of the phylogeny (Figure 2). For example, reaction R01087 is catalysed by Maleate cis-trans-isomerase which is utilized in maleate assimilating and hightemperature bacteria. A second sub-network involving both ADP-forming acetate and propanoate CoA ligases is potentially used in the organisms to convert between acetate and propanoate and their corresponding CoA forms.

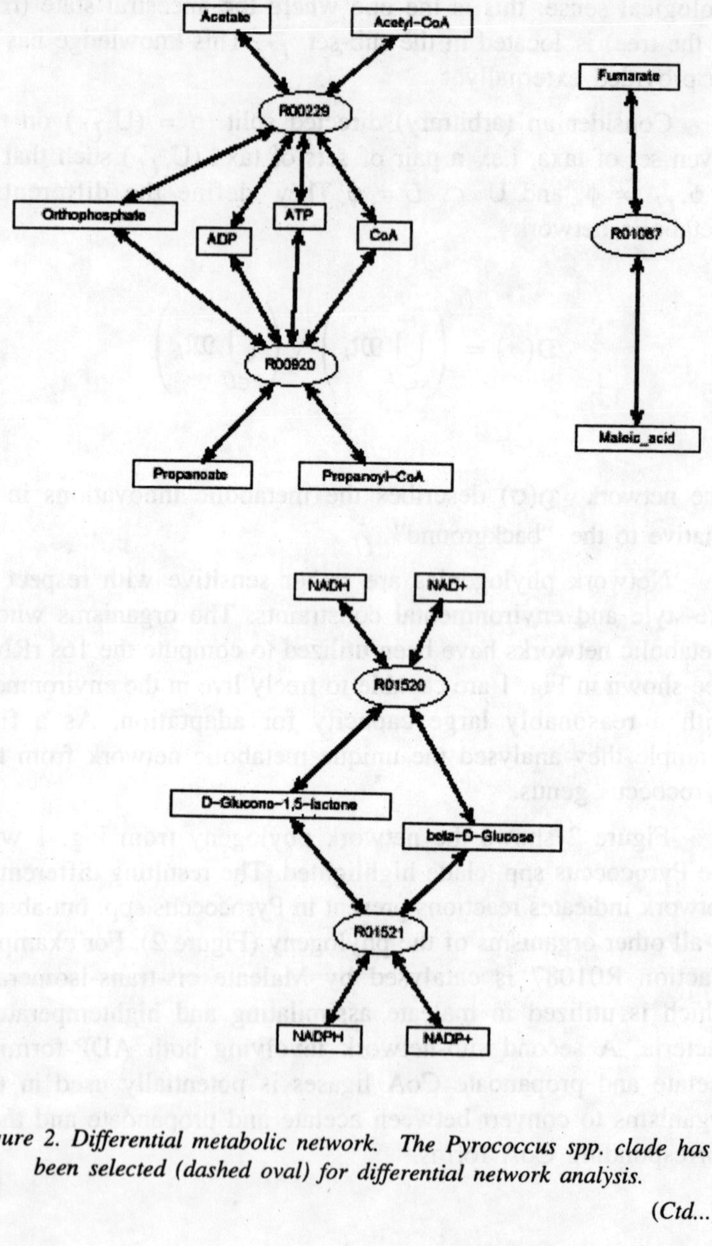

Figure 2. Differential metabolic network. The Pyrococcus spp. clade has been selected (dashed oval) for differential network analysis.

(*Ctd...*)

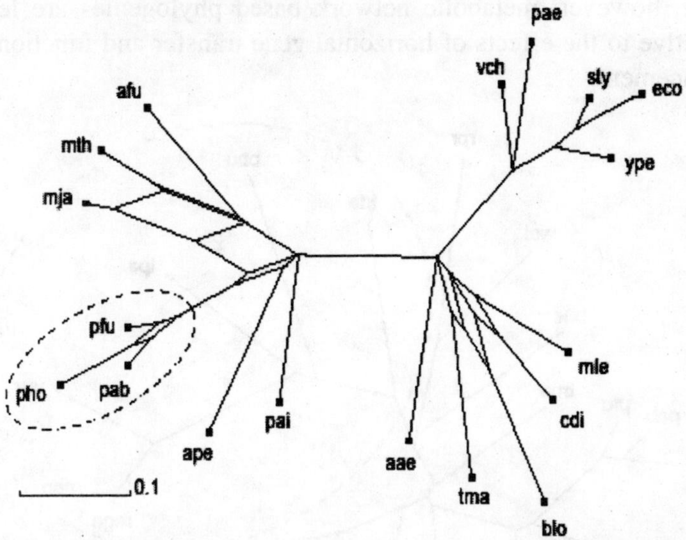

Figure 2. Differential metabolic network. The Pyrococcus spp. clade has been selected (dashed oval) for differential network analysis.

As a second example they analysed their set of reference organism (Figure 1) with obligatory intracellular pathogens. Figure 3 shows the phylogeny with the selected pathogens (dashed oval). Interestingly, Mollicutes, such as My-coplasmae and Spirochaetes, such as Treponema are grouped together. They all possess a minimal gene-set, and thus a highly optimised and host-dependent metabolic network. Surprisingly, this set of organisms has specific reactions that are absent in the remaining organisms of the phylogeny.

The Vienna Reaction Network Library introduced above treats chemical reaction networks, and metabolic networks in particular, as directed hypergraphs. A framework borrowed from set algebra provides natural definitions of unions, intersections, and differences that can be used to compare the metabolic networks of difference organisms. Metabolic networks convey phylogenetic information and can indeed be used to infer phylogenetic relationships of free-living organisms in a way that is similar to gene-content based approaches. In contrast to the latter, however, metabolic network based phylogenies are less sensitive to the effects of horizontal gene transfer and functional replacement.

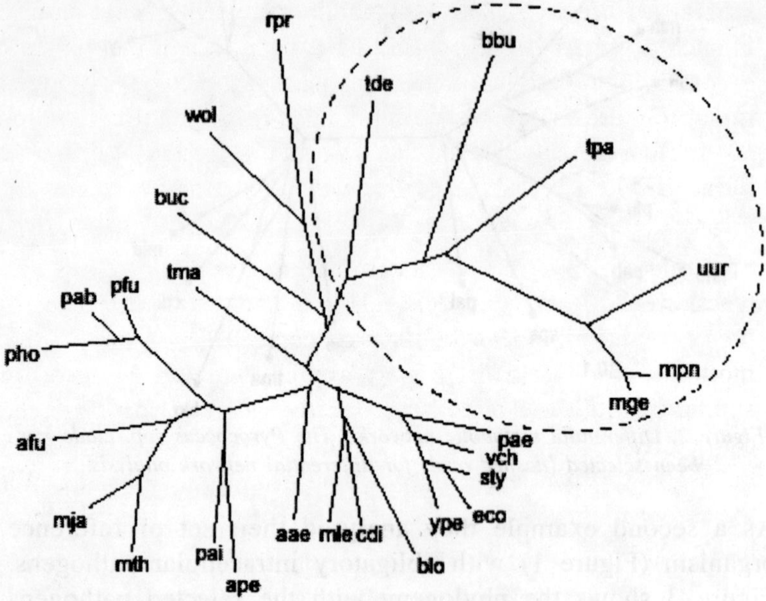

Figure 3. Unrooted network phylogeny using PHYLIP with the Fitch-Margoliash al-gorithm. A set of obligatory intracellular pathogens has been selected (dashed oval) for differential network analysis.

Differences of metabolic networks among subtrees of a trusted phylogeny, or more generally, along any split of interest in a set of organisms can be computed directly, making it easy to study metabolic innovations in particular clades. A first application of their network phylogeny analysis involved three members of the Pyrococcus spp. clade.

The metabolic reactions resulting from the split between the Pyrococci and the remaining organisms involve the maleate cis-trans-isomerase reaction, ADP-forming acetate and propanoate CoA ligase reactions as well as beta-D-Glucose:NAD(P)+ 1-oxoreductase. Their second example considers a class of intracellular pathogens that includes Mycoplasmae, Ureaplasmae, and Spirochete.

Their restricted repertoire of metabolic reactions reflect the specialised life-style. Many metabolic pathways are not required in such a rich environment and have been lost in the course of evolution. On the other hand, constructing a network phylogeny including these microbes, they observe metabolic reactions assembling an unconnected network that is present in this set of intracellular pathogens and absent in remaining organism.

Such reactions include phosphorilisation and conversions of sugars and derivates, deaminating lyase reactions, and reactions involving carnitine, choline and tetrahydrofolate. At present, metabolic network data are compiled by a multitude of methods, and at least in part are constructed by genomic similarity with other organisms. Strictly speaking, therefore, they cannot view metabolic network data such as those complied in the KEGG database as independent from genomic data. With the recent advances of experimental techniques in metabolomics, however, the situation is rapidly improving.

Their comparative approach to metabolic network analysis, which focuses on individual reactions rather than on aggregate feature such as pathways, simplifies the identification of metabolic innovations and, in particular, facilitates the recognition of organisms as potential biological threat agents based on their metabolic repertoire. Furthermore, the ability to easily identify differences in metabolic capacity between pathogens should be useful towards a refined classification of pathogenicity based on metabolic capabilities.

17

REGULATION OF METABOLIC PATHWAYS

The wiring of a metabolic network is represented by a static set of chemical reactions. Most of them are well characterized and could be seen as the 'hardware' component of metabolism. Much less is understood about its 'software' component—the complex set of regulatory strategies used by the cell to perform a variety of tasks in different environmental conditions. Mastering the language of this software would not only constitute a major intellectual achievement, but could also lead to many practical applications, such as making the application of costly metabolic processes more efficient.

Although the transcriptional and non-transcriptional regulation of metabolic flow in single pathways has been studied extensively, the discovery of recurring regulatory patterns and modular organization in the transcriptional regulatory programme of metabolism remains an ongoing challenge in functional genomics. The availability of pathway databases and large expression datasets offers new approaches for reaching this goal, as explored by Ihmels et al. for the metabolic network of Saccharromyces cerevisiae. Their systematic approach could help bridge the gap between top-down data analysis and bottom-up model building.

Ihmels et al. began their study by asking whether all genes within a given pathway (defined using the Kyoto Encyclopedia of Genes and Genomes database) are, as might be expected, co-regulated. It appears that, for most pathways, only a subset of genes displays significant co-expression (e.g. 24 out of 46 genes in glycolysis). The Signature Algorithm is used to identify systematically any classes of genes that are associated with a given metabolic pathway and whose expression is correlated under a certain set of conditions. These sets of genes, or transcriptional modules, consist mainly of linear arrangements of enzymes in the metabolic network. These modules could also include genes from feeder pathways, transporters and transcription factors.

The large-scale analysis of enzyme co-regulation patterns has revealed an arrangement of functional associations that crosses the boundaries of classically defined pathways. Therefore, a 'dromocentric' view of metabolism (from the Greek word dromos, meaning pathway) should be complemented by a 'modulocentric' approach, in which modules rather than conventional pathways are the elementary functional units. Grouping of genes by modules rather than by pathways might also provide cleaner and more significant signals when analyzing new metabolic gene expression data.

The cellular response to changing conditions often involves altering metabolic fluxes at crucial branching points by using a combination of allosteric, covalent and transcriptional regulation. How important is the role of gene co-regulation at these junctions? At diverging branches, co-regulation of the incoming reaction and one of the two outgoing reactions is prevalent. This pattern is suggested to confirm the preference for linear flows along pathways.

Although coordinated regulation of enzymes along a linear pathway is a documented and fundamental modus operandi, it should not be expected to be ubiquitous and to necessarily imply linearity of the metabolic flow. In some cases, simultaneous flow through different branches is necessary, for example, for the allocation of precursors in the synthesis of essential biomass components. In other cases, branching pathways might truly represent differentially regulated alternatives.

The recurring expression patterns found by Ihmels et al. also suggest the existence of general mechanisms for the dynamic control of fluxes. Additional insight into the dynamics associated with different co-expression patterns would require a more systematic analysis of all regulatory strategies at branch points, possibly using metabolic control analysis, biochemical systems theory or flux-oriented theory.

Isoenzyme switches

For several metabolic reactions, the catalyst can be one of two (or more) distinct enzymes (isoenzymes). One possible role of isoenzymes could be either the redundancy or enhancement of catalytic function. In such cases, one might expect isoenzymes to be co-regulated. Ihmels et al., intrigued by the observation that isoenzymes often correlate with distinct modules, systematically analyzed isoenzyme co-expression patterns. In the majority of the examined pairs of isoenzymes (63%), they found that the two enzymes are not co-regulated.

Rather, each of them (or at least one of them) correlates with a distinct enzyme that is an immediate neighbor in the metabolic network. This result indicates that different isoenzymes are frequently involved in distinct cellular responses that are controlled like switch mechanisms. For example, the two hexokinase isoenzymes HXK1 and HXK2 correlate with glycogen storage and glycolysis, respectively. It had also been proposed that this kind of specialized isoenzyme could be mediated by enzyme complex formation and channeling.

Interestingly, some pairs of isoenzymes might undergo both transcriptional and non-transcriptional regulation. For example, the glutamate dehydrogenases GDH1 and GDH3—which correlate with different neighbor reactions—were also found to fulfill distinct tasks under allosteric control. A switch in isoenzymes usage could be induced by unanticipated cellular requirements. Yeast grown in the presence of Cd^{2+} produces large quantities of glutathione for detoxification, which increases the need for sulfur. This requirement triggers the replacement of sulfur-rich isoenzymes with sulfur-depleted equivalents. This transcriptional switch, especially evident in the ENO1-ENO2 and PDC1-PDC6 isoenzyme pairs, was not detected in the analysis by Ihmels et al. This serves as a reminder that the variety of conditions encountered during evolution might cover a much larger range of possibilities than those that are currently sampled experimentally. Conversely, this example suggests that some of the observed isoenzyme switches could be associated with other responses to amino acid shortage.

Regulatory Software

At a whole-cell level, the hierarchical structure of metabolic functional modules can be inferred using an 'unsupervised' variant of the Signature Algorithm—the Iterative Signature Algorithm. By continuously decreasing a co-regulation threshold parameter, small modules merge into larger modules with less specific functions and decreasing levels of co-expression.

Only three modules are found at low resolution: stress, protein synthesis andamino acid biosynthesis.Particularly striking is the division of these two lastmodules, possibly reflecting a distinction between fast growth under amino acid-rich media and slow growth under amino acid-poor media, respectively. The classification proposed by Ihmels and colleagues focuses on the functional roles of the different genes, and is shown to differ from a previous analysis that was based on network topology.

For example, fermentable and nonfermentable carbohydrate degradation pathways are far apart in the function-based hierarchy, despite their proximity in a topology-based classification. A different quantitative analysis of the relationship between gene regulation and network topology has recently been proposed. They found that, on average, the closer two genes are in the metabolic network, the higher their correlation in expression. Above a characteristic pathway length, genes tendnot to be significantly coregulated. Altogether, these analyses of the connection between network topology and gene expression constitute new examples of data integration, which might be extended to include protein levels or other data.

BIBLIOGRAPHY

Albert R., and A.-L. Barabasi. "Statistical mechanics of complex networks". *Rev. Mod. Phys.* 74, 47–97 (2002).

Jeong, H., Tombor, B., Albert, R., Oltvai, Z., Barabasi, A., "The large-scale organization of metabolic networks. Nature" 406 (2000)

Kim, B., Yoon, C., Han, S., Jeong, H., *Path nding in scale-free networks*, (2002)

Kohn K. W., "Molecular interaction map of the mammalian cell cylce control and DNA repair systems". *Mol. Biol. Cell* 10, 2703-2734 (1999).

Laurent M., and N. Kellershohn "Multistability: a major means of differentiation and evolution in biological systems". *Trends Biochem. Sci.* 24, 418–422 (1999).

Minoru, K., Susumu, G., Shuichi, K., Akihiro, N.: "The KEGG databases at GenomeNet. Nucleic Acids Research" 30(1) (2002)

R. Alves, R.A. Chaleil, M.S., "Evolution of enzymes in metabolism: a network perspective", *Journal of Molecular Biology* 320 (2002)

Salazar I., J. Garcia-Fernandez and R. V. Sol´e "Gene networks capable of pattern formation: from induction to reaction-diffusion". *J. Theor. Biol.* 205, 587–603 (2000).

Smolen, P., D. A. Baxter and J. H. Byrne "Mathematical modeling of gene networks". *Neuron* 26, 567–580 (2000).

Thieffry D., A. M. Huerta, E. P´erez-Rueda, and J. Collado-Vives. "From specific gene regulation to genomic networks: a global analysis of transcriptional regulation in *escherichia coli*". *BioEssays* 20, 433–440 (1998).

van Helden, J., Naim, A., Mancuso, R., Eldridge, M., Wernisch, L., Gilbert, D., Wodak, S., "Representing and analyzing molecular and cellular function using the computer", *Journal of Biological Chemistry* 381(9-10) (2000)

Vogelstein, B., D. Lane, and A. J. Levine. "Surfing the p53 network". *Nature* 408, 307–310 (2000).

Wagner A., "Robustness against mutations in genetic networks of yeast", *Nature Genet.* 24, 355–361 (2000).

Watts D. J., and S. H. Strogatz. "Collective dynamics of 'small-world' networks". *Nature* 393, 440–442 (1998).

Williams R. J., ans N. D. Martinez. "Simple rules yield complex food webs". *Nature* 404, 180–183 (2000).

INDEX